BURN FUEL BETTER

FROM **HELPLESS** TO **HOPEFUL**
IN THE RACE AGAINST CLIMATE CHANGE

A book by Donald Owens

Published by
BEVERLY HILLS PUBLISHING
468 Camden Drive
Beverly Hills, California 90210

Beverly Hills Publishing Inc.
www.beverlyhillspublishing.com

ISBN: 978-1-7360900-7-7

Set in Minion Pro

"For someone visiting Earth for the first time, the real treasures here would all be free. The smell of a sunlit prairie, the taste of a cold cup of spring water, the crunch of trackless snow underfoot—these are some of the earth's supreme treasures. On intergalactic maps, if there are such things, the place where we live must surely be designated as a magical garden in space, a place of astounding beauty."—Steve Van Matre

Table of Contents

INTRODUCTION

From Helpless to Hopeful

"Only within the moment of time represented by the present century has one species—man—acquired significant power to alter the nature of the world."—Rachel Carson

Let me tell you something about human beings (just in case you didn't know).

Humans are miraculous, complex, masterfully creative beings who are uniquely capable of actualizing the most magnificent and the most absurd feats of the imagination. If we are able to think of something, even for a split second, we are equally able to find a way (or multiple ways) of transforming that thought into a tangible reality. The absolute infinite universe of things that have been planted into the minds of human beings cannot and should not be underestimated. From the development of modern-day wingsuits, when someone somewhere realized that an increase in the surface area of the human body would enable a significant increase in lift that would allow one to fly like a bird, if they decided to jump off of the side of an extremely high cliff, to the first recipe for a key lime pie (my personal favorite)—there are truly no bounds to what the human imagination is capable of dreaming up and actualizing.

Since the earliest recorded moments in history, humans have been curiously tinkering within the physical realm, bending and blending the elements together, harnessing the wind and the rain, seeking to understand, measure, manipulate, and ultimately, dominate nature.

For example, if we look back to the first century B.C., we see the ancient compulsion to transform and control matter successfully evidenced in the alchemists. It is that same relentless curiosity that enabled us to wrangle electrical charges to light and heat our homes, that also led us to construct artificial wave pools to play upon, and that inspired us to send men and machines to the moon and beyond. The scenes that were once portrayed as science fiction on my television screen as a boy are now actualized and played out in (and across) the various devices that my children and I use each day.

Every time we ask Alexa to tell us the time, trust Siri to direct us to our next destination, or enable our phone's voice dictation feature to FaceTime our family members at only the proper moments, we are interacting with the kind of artificial intelligence that was once merely fodder for futuristic fantasies. (FYI, for some reason, I rarely speak to either one of them.) And, when a viral pandemic plagued our planet and forced entire populations to scatter, separate, and stay home for months on end, we were still able to meet virtually to cook dinner with our families, choreograph ballets, and sing sweet birthday serenades from the comfort of our kitchens and couches.

I am always in awe of humanity and our infinite capacity to create. I still don't know what possessed Galileo Galilei to build a telescope sophisticated and strong enough to enable his discovery of Saturn's rings almost four hundred years ago. I still don't know how magicians, past and present, are inspired to craft such unbelievable, mind-bending sleight of hand illusions, nor do I understand how, from the beginning of time, musicians have been able to create the most incredible combinations of sounds with instruments and voices that can soothe the human spirit. With each passing minute, I am constantly reminded of the boundless, infinite abilities of human beings.

But we definitely have our downsides, too, don't we? Some of the things we do on this planet can be pretty messed up, even if some of those things were originally well-intentioned. The moment humankind evolved beyond that which can be found or sourced naturally within our world and constructed entirely new universes of imagination to experience and explore, we also created the source (and, consequently, became the key driver) of our own ultimate demise—which will stem from the mass infiltration of poisonous toxins and harmful elements that our endless "advances" have unleashed on our atmosphere. These toxins (namely CO2, plastics, particulate matter and, most notably, black carbon) will surely, sooner than later, transform our native planet completely, and not in a good way.

Left unattended, humankind's relentless consumption of natural resources, along with our consumer-driven demand for fossil-fuel-produced goods and services, will devastate our home and all of its inhabitants beyond recognition.

We've got a problem.

We are rapidly wearing out our welcome here.

And we are all at fault—every single one of us—even the most committed environmentalists, and even those among us who may not believe that climate change is a real thing.

Still, the fact that we are having an adverse effect on the planet in multiple ways is becoming harder and harder to deny, especially when we see over three hundred million tons of plastic being produced every year for use in a wide variety of extremely useful applications, and at least eight million tons of that plastic ends up in our oceans. Plastic pollution alone makes up a whopping 80 percent of all marine debris, from surface waters to deep-sea sediments. The production of all of that plastic is tied, both indirectly and directly, to the burning of fossil fuels—from power plants to combustion engines. All of these combustion engines, but especially those that are powered by diesel fuel, produce particulate matter and black carbon, which are coating our ice shelves (causing them to melt) and lining our lungs (causing us to die).

I believe we are collectively becoming more aware of the plastics issue and are genuinely trying our best to get a handle on our production, consumption, recycling, and repurposing of it. But we are ***not*** currently doing anything about particulate matter and black carbon. Nothing big enough to make a lasting impact, anyway, since particulate matter and black carbon have proven to be positively devastating to human health, and as will be later explained, equally as devastating to the planet. Hopefully, even if you don't like the concept of climate change on the planet, you do still like the concept of "human life on earth."

For those of us who do believe in climate change, most of us attribute the rapid innovation, modernization, and globalization of the twentieth and twenty-first centuries—the eras of production, convenience, consumption, and quick gratification—as the biggest catalysts to climate change. After all, the past 121 years contributed mass-scale factories, hyper-armed militias, agricultural machinery to feed the world, consumer vehicles, worldwide container shipping, unbelievably effective communications, and fast fashion to the mix. As a result, many of us also have heard that CO2 is the leading and most important cause of climate change (it's really not) and that converting to green energy will save us from it (it will, but it won't quickly enough).

The funny thing is, it really doesn't matter. We're all wrong.

For decades we have all been pointing fingers past ***the true culprit of carbon crimes, which is our own unending demand (not request) for the lifestyle comforts we've come to expect.*** Every single one of those comforts—whether it be heating our homes, driving our cars to work, picking up a package from the grocery store or the post office, or brushing our teeth with clean water in the morning—all cause the burning of fossil fuels that produces ***black carbon, which is about 1,500 times worse than CO2 emissions when it comes to climate change.***

Now, you don't need to take my word for it. That's the beauty (and the horror) of the almost instant knowledge of the Internet. Several more notable and reputable sources than I have said the very same thing for some time now. Take the Climate & Clean Air Coalition (CCAC), for instance, which released a report with a handy infographic revealing that "black carbon has a warming impact on [the] climate [that is] 460–1,500 times stronger than CO2 per unit of mass."[1] That report also goes on to reveal through an infographic how "household cooking and heating account for 58 percent of global black carbon emissions." Smaller-scale transportation contributes the next lion's share of emissions at about 26 percent.[1] This includes one of the three vehicles your neighbor has parked in his/her garage, the school buses that take our children safely to school each weekday, and the big rig that's delivering the newest batch of Amazon packages you ordered while snacking the other night and had to have shipped from two states over.

It's worth noting that that very same, very official infographic shows that the most famous environmental offenders, which are fossil fuel operations and waste production processes, only collectively account for a whopping 8 percent of total black carbon emissions. Which, again, are up to 1,500 times more capable of warming our planet than CO2.

Historically, we've totally missed the point. Until now.

People are starting to wake up.

That said, even those with some serious skin in the game don't fully understand the scope, the source, or the solution of the problem. Take those feisty, impassioned environmentalists (whom I love, by the way, because in my heart I am one of them), for example, who have dedicated their lives to picketing fossil fuel companies and calling for a boycott of "Big Oil" for its presumed role in accelerating climate change. Those environmentalists are actually, in truth, equally culpable for the climate crisis we find ourselves in today. Because, I guarantee you, those same folks are still going to turn on their home's central heating in the face of an unanticipated and unprecedented Texas blizzard, if for no other reason than to survive through the frosted night. Those same people are going to bathe their babies in warm, clean bubble baths at night, so as to prevent them from catching a chill or a cold from the water. They are still going to frequent all types of restaurants that, to varying but unavoidable degrees, rely on the material luxuries of heat, plumbing, manufactured equipment, refined products, shipped produce, and chilled freezers. They are still going to use cell phones and the Internet to communicate with one another. Despite all of their better efforts, they are still going to exist in an industrialized world, where harmful black carbon created as a by-product of burning fossil fuels will forever be produced and released into the atmosphere so long as we continue building, manufacturing, transporting, sorting, storing, and heating or cooling any one of our modern comforts, ***which we will never stop doing***. Unless we want to revert back to preindustrial society with no cell phones or Internet or indoor temperature control, or be thrust forward into a postapocalyptic society (in which nothing new is ever created and everything is scarce), we will never be able to hide from our own demand for comfort and its consequential, potentially life-ending pollution. Unless we want to start rocketing ourselves and our loved ones up into the stars in an attempt to develop civilization on other planets, we cannot ignore this very real truth.

Please, if you take even one thing away from this book, let it be this singular truth: ***Black carbon, which is created by absolutely everything that burns fuel, is much, much worse for our environment than CO2 (1,500 times worse) because it is released into the air and falls onto our glaciers, coating them and causing them to absorb light and heat instead of reflecting that light and heat, thereby causing them to melt. Period. PERIOD!***

And yet, nobody seems to be paying too close attention to black carbon's role in our rapidly declining climate. However, don't despair; there's a reason for that. It's because, as a society, we simply didn't know about it (until now). And, in the face of such lack of knowledge, many of us feel helpless in the fight against climate change or global warming as it was previously referred to. I mean, it's human nature. When there doesn't seem to be an easily identifiable, highly implementable solution to a global problem (and at such an unimaginable scale), it is easy to feel small and insignificant in our efforts. It's easier to forget the facts, or to simply fail to ask the right questions in the first place.

It is also human nature to find a way to survive against all odds. To battle every great illness and enemy and catastrophe with the very best we've got, and to keep on fighting for our lives and the lives of our loved ones. If we seek to become masters of our fate, as many of the master philosophers would challenge us, we must first become masters of our fortress—and take responsibility, action, and collaborative steps forward if we have any hope of continuing the chapter we've started here on earth.

The good news is: The remedy for climate change is here, so you can cross that one off of your list. Check, check. Inhale, exhale. We can move slowly from feeling helpless to hopeful, because ***now we can cut black carbon emissions from anything and everything that burns fuel (for now combustion engines only) right at the source BY UP TO HALF.*** (Granted, "half" is not quite as great as "all," but it is a start and can help buy humanity some much-needed time in the race against climate change.)

We've developed something, which we'll dive into more deeply later, that can work with every existing fossil-fuel-burning combustion engine to reduce particulate matter production at the source, before it even has a chance to develop and disperse into the atmosphere—while simultaneously producing oxygen for the environment. This is really, really, really good news. This is a big win, and one that can turn the tide in our favor, or at least give us some hope. I'm hoping that this development can take away the feelings of helplessness regarding climate change out of our collective minds, just a little bit. That it can spark newer and even better iterations of our solution for wider adoption and application. That it can maybe brighten the climate change conversation forever (or, at least, for a little while).

In fact, the whole reason I'm writing this book now, other than to highlight an almost impossibly simple yet highly effective solution, is with the intention that this development, this information, can bring hope to what has long been a feeling of hopelessness regarding climate change for so many. To bring hopefulness to those who feel helpless in the face of the fate of the planet (which, honestly, is all of us). To focus on the future, rather than to despair in the present. It's possible, now. We have the answer right here in our hands.

But even this should not be viewed as a panacea.

This is an almost inconceivable undertaking, which will require developments, iterations, and adaptations for an almost unlimited number of applications. That said, while this feat is much bigger than any one of us individually, it is not bigger than all of us collectively. And I hardly expect to do this on my own. No one could ever do it alone. I need people like you, with your own individual kaleidoscope of talents and creativity, to partner with the rest of us in this pursuit. I also don't anticipate that people will act immediately on my advice just because I say so. Of course not. I'm not a famous chemical engineer or lauded environmental scientist or award-winning inventor or esteemed political activist. I'm just a lifelong tinkerer with a few different degrees who is blessed with a compulsive curiosity and a tendency to find simplified solutions. Please note, I didn't choose this. I didn't seek out a solution to climate change. I didn't even know what particulate matter and black carbon were before I started testing our technology at the lab and discovered that we cut it in half. I literally turned to the lab technician after he revealed this finding to me and said, "Oh really, we cut particulate matter by almost 50 percent? Cool. Just out of curiosity, what is particulate matter?" And it wasn't until years later that I discovered that black carbon is a component of particulate matter. This climate change solution found me, and once it revealed itself to me, I felt compelled to shout it from the rooftops, in the hopes that some of you incredible, limitless, brilliantly talented human beings will collectively conclude that black carbon is, indeed, our true enemy and collaboratively decide to do something about it so that one day, those rooftops and glaciers won't be speckled with the dark dust of black carbon—and the sun will still be shining just as warm as it should be.

Thank you for joining me on this journey.

Now … let's go save the world.

CHAPTER ONE

Black Carbon: Enemy of the People (and the Planet)

"The opportunity of defeating the enemy is provided by the enemy himself."—Sun Tzu

If you live, work, and breathe in the twenty-first century, chances are—you are aware that the planet is warming, the ecosystem is moving in the wrong direction, and humankind is scrambling to find a way to clean up after itself. It's as though we have been throwing a wild, nonstop party at our parents' house while they were away on vacation for the past two or three centuries, to the point where the place has become unrecognizable and almost unsuited for ordinary life, and we just got word that they'll be returning home tomorrow. Except, in this case, the lawn has been uprooted, the kitchen sink is flooding over ... and the house is on fire. Whether you fall within the camp that adamantly denies that climate change exists (it does) or the one on the opposite end of the spectrum, where you have dedicated every waking hour of your life attempting to spread awareness and spark change to reverse global warming while we still can, you cannot avoid the sharp reality—our time on earth is ticking if we do not act now.

The topic of climate change has gained momentum over recent decades in direct proportion to the rising visible, undeniable, and dangerous manifestations of a rapidly warming planet. We can see the evidence of our environmental impact all around us, even if we would really rather not pay attention to it. Take just the past eighteen months or so, for example. When the great state of Texas can be brought to its knees in less than four days by one swift gut punch from Mother Nature, like it did earlier this year when an unprecedented winter storm left millions powerless, heatless and, in many cases, without any food or water, you know something must be amiss in the atmosphere. (Sure, Texas is familiar with how to handle tornadoes and flash flooding—but single-digit snowstorms interrupting Valentine's Day plans? That's a head-scratcher.)

Or look at California, a richly variegated coastal landscape of forest, snow, sea and desert, which suffered its worst wildfire season on modern record in 2020. It has been estimated that, by the end of that year, more than 4 percent of California's land (roughly 4.3 million acres of 100 million total acres) had been charred by nearly ten thousand separate fires.[2] Now, California is no stranger to devastating natural disasters, ranging from earthquakes to landslides to wildfires, but never to this degree. When the seaside, sumptuous town of Malibu is suddenly engulfed by devilish, fast-moving walls of fire—forcing even the most extravagantly rich households to abandon their estates and flee for safety—and a stark, ash-laden red sky settles over most of the massive state for more than a week, it is clear that things are not as they should be. California's nightmarish 2020 fire season accounted for about 40 percent of the record-setting year's 10.3 million acres of total land burned on American soil alone.[3] If we want to expand our lens to a more global view, we can turn to the even more disastrous bush and forest fires that ravaged Australia and the Amazon[4] forest that same year. These caused a staggering loss of approximately 7.5 million acres of land and more than four billion animals in those areas, some of which are now on the path toward extinction.[5] Add to this equation the bleaching of the Great Barrier Reef and the "Manhattan-sized" (twenty times over) iceberg that just broke off of the Antarctic shelf earlier this year, and we no longer have a recipe for disaster—we are smack dab in the middle of the disastrous feast.

That said, it's not an unexpected feast. We have been preparing for this climatic chaos for decades, pushing green energy agendas, picketing fossil fuel usage, and pivoting toward more sustainable behaviors, but unfortunately, it seems that we have been putting much (if not all) of our focus on the wrong adversaries. Most environmental talks have targeted the mass accumulation of harmful greenhouse gases, particularly carbon dioxide (CO2), in the atmosphere, along with major industrial players, like "Big Oil" companies (more on that logical fallacy later), as the key culprits of climate change.

Despite the fact that, yes, atmospheric CO2 levels are now at the highest point they've been in almost a million years (nearly double that which was recorded about seven hundred thousand years ago during a prehistoric, interglacial warm period), and much of that steep increase is linked directly to industrialization and the rapid-scale burning of fossil fuels (again, we'll dive more into this topic in later chapters)—neither CO2 nor the Big Oil industry will be the sources of an immediate solution to stem global warming.[6]

The real enemy of the people, and the planet, is an insidious solid waste known as black carbon, which, according to the Climate & Clean Air Coalition, is up to 1,500 times worse than CO2 in terms of its capacity to warm the planet.

Why, then, is nobody really talking about tackling black carbon? The reason is that, as a society, we just don't know about it.

Now, we are. Welcome to the conversation.

At the risk of being redundant, I have chosen to reiterate and recontextualize one crucial truth throughout the course of this book, because it is the very core and crux of this issue: ***"Black carbon ... coats the glaciers ... causes them to absorb light and heat instead of reflecting light and heat ... and they melt ... Period."*** Please scan the QR code if you'd like to see some of its effects.

One of the reasons I am repeating this sentiment over and over again is because, a few years back, I didn't know what black carbon was or how it affects us. Like so many others, I spent the lion's share of my life totally unaware of how this sooty terror has slowly been smothering our planet.

I was only made aware of black carbon as a tangential result of some lab tests I had been running on a fuel-saving device that I'd engineered in my garage—a device that, at the time, had no planned or even foreseeable relationship to ***anything*** emissions related. Frankly, while developing a prototype for that device, I was more fixated on fuel and cost savings than on reducing pollution, and I was having a heck of a time getting the thing to do that (you can flip to the end of the book for more details about that masterclass in trial and error). So, when the technician at the engine testing lab where I'd been conducting these endless prototype experiments informed me (after a particularly disappointing fuel savings test) that the results report revealed that my device had apparently "cut particulate matter production in half," I did what any self-respecting scientist/engineer would do—I acted cool, nodded my head, and then asked him: "What the heck is particulate matter?"

Turns out, particulate matter (PM) is a combination of microscopic liquid and solid waste particles that can be created naturally but often are formed as a by-product of man-made, fossil-fuel-burning machines and engines. ***In essence, particulate matter is the unburned fuel that occurs during fuel combustion.*** When emissions result from the burning of fossil fuels in engines, particulate matter becomes a pollutant that is positively poisonous to humans. According to the Environmental Protection Agency (EPA):

> *"The size of [PM] particles is directly linked to their potential for causing health problems. Small particles less than 10 micrometers in diameter pose the greatest problems, because they can get deep into your lungs, and some may even get into your bloodstream. Exposure to such particles can affect both your lungs and your heart. Numerous scientific studies have linked particle pollution exposure to a variety of problems, including: premature death in people with heart or lung disease; nonfatal heart attacks; irregular heartbeat; aggravated asthma; decreased lung function; and*

> *increased respiratory symptoms, such as irritation of the airways, coughing or difficulty breathing."*[7]

Beyond these more established and understood respiratory and cardiovascular side effects of particulate matter inhalation, recent studies have found that high levels of PM pollution "may damage children's cognitive abilities, increase adults' risk of cognitive decline, and possibly even contribute to depression."[8] And the more we look into the harmful effects that particulate matter has on humans, the deadlier we discover it is. According to the World Health Organization (WHO) and a recent study released this year from Harvard University:

> *"More than 8 million people died in 2018 [alone] from fossil fuel pollution [and outdoor airborne particulate matter], significantly higher than previous research suggested ... Researchers estimated that exposure to particulate matter from fossil fuel emissions accounted for 18% of total global deaths in 2018—a little less than 1 out of 5."*[9]

This, of course, doesn't account for the higher morbidity rates from COVID-19 that have been linked to particulate matter exposure and the underlying respiratory conditions that are caused by it. According to another study recently released by Harvard University T.H. Chan School of Public Health that sought to understand if a correlation exists between exposure to particulate matter and complications (including death and other serious illnesses) from the novel coronavirus, "a small increase in long-term exposure to PM2.5 leads to a large increase in the COVID-19 death rate."[10]

While reviewing the November study's findings for *The New York Times*, environmental policy reporter Lisa Friedman wrote, "The paper found that if Manhattan had lowered its average particulate matter level by just a single unit, or one microgram per cubic meter, over the past 20 years, the borough would most likely have seen 248 fewer COVID-19 deaths by this point in the outbreak."[11] It's safe to say that particulate matter exposure is a matter of life and death for humans.

But that's just what it does to the bodies of living beings here on earth. If we also look at the effect these airborne particles have on our ecosystems and environment, we see that particulate matter can have a myriad of harmful impacts on our land, air and water, including but not limited to significantly decreasing visibility and air quality; altering and depleting the natural nutrients found in soil; damaging forests, farms and crops; and even turning bodies of water and rain into acid. It is absolutely nasty stuff, and it is everywhere.

Now, the true beast of this deadly particulate matter category is black carbon. Black carbon, the major component of fine particulate air pollution (PM2.5), is, essentially, defined as tiny fragments of unburned fuel created by incomplete combustion and released out of diesel and gas engines into the air and onto surfaces. Often colloquially referred to as "soot," black carbon can be seen by the naked eye as a solid, dark toxic dust that falls and accumulates, without plan or prejudice, all around us. If you live near a port city, you probably see it all the time, when you remove it from your outdoor furniture.

"Black carbon is the material that burns in an orange flame," explains University of Illinois atmospheric sciences affiliate professor Tami Bond. "What you see in fire is black carbon glowing." Through the burn, Bond adds, accumulated particles of almost pure carbon (several thousand times smaller than the size of a human hair) are released into the air from the flame.[12] Oftentimes, while fine PM particles are airborne, they can be inhaled and introduced into the respiratory and circulatory systems of every kind of life form on earth, from birds to fish to humans.

Because of their size and mixability with other agents of transportation, black carbon particles can travel on the wind across great continental distances toward more vulnerable ecosystems, like Antarctica. Once created, black carbon particles attach to other elements and materials rising from the ground, such as dust and sulfates, which rise together through the atmosphere and are introduced into the water cycle.

While these black carbon compounds will naturally circle in and out of the atmosphere by ascending into clouds and falling back to the earth through rain and snow, their harmful absorptive qualities remain intact, and therefore their ability to directly heat the surfaces they fall onto makes them just as dangerous on land as they are in the air.

You see, due to its dark, dense color, black carbon does an excellent job of absorbing sunlight (in every single wavelength) as heat, rather than bouncing that light off of itself, and then transfers that heat directly into the atmosphere. While the heat from black carbon will have a much shorter half-life in the ozone than that produced by CO2, it is much, much more effective in traveling long distances, depositing itself into/onto disparate ecosystems, and rapidly destroying those resources.

While we've seen the majority of estimates placing black carbon's warming potential at about 1,500 times that of CO2, one study published by the National Center for Biotechnology Information (NCBI) suggests that these particles have "roughly a million times the heat-trapping power of CO2."[13] The study goes on to state that, when black carbon "falls out with precipitation on snowpack or ice, it absorbs heat and accelerates melting by interfering with how those white surfaces reflect sunlight back to space." This is especially problematic, as we've seen massive accumulations of black carbon visibly lacing and lining the polar ice caps—where it is effectively cooking them from the top down. Remember that more-massive-than-Manhattan chunk of ice that just broke off of the Antarctic ice shelf we mentioned earlier? Yeah. That was a direct consequence of its black carbon content.

Here's the real kicker: Black carbon is created by anything and everything that burns fossil fuels. ***READ THAT AGAIN. Everything (and I mean everything) on earth that burns fuel creates and releases black carbon into our environment.*** More on that later.

And, while black carbon is also released from every single chimney, gas oven, and cookstove across the globe (with poorer developing countries actually contributing a large hunk of that load through their necessary, widespread use of individual cookers and old diesel engines to survive), we will be focusing predominantly on the black carbon creation that stems from the burning of diesel and other like fuels globally for our purposes here in this book. The truth is, all industrialized countries are key contributors to the dark, absorptive soot that falls directly onto our glaciers, thereby making each of us a huge, addressable patron and participant of global warming activities.

We need to do better.

We need to stop the issue at the source, and fast.

Now, maybe you're thinking to yourself, "That's great and all, Don, but what about the existing efforts to pull harmful carbon out of the atmosphere and store it? I read an article about that recently. Surely that's going to help remedy this issue." You're right; carbon capture is a magnificent feat of science and imagination that will be an asset to the fight against climate change. But before I even begin to discuss the limitations of carbon capture, one thing needs to be widely understood. ***Black Carbon is not captured by carbon capture and black carbon is not CO2. It is 1,500 times worse than CO2.***

Going back to carbon capture, there are still several inherent kinks to this process that must be sorted out, namely (a) its scalability, (b) its mitigation burden, and (c) its cost to implement. As David Roberts, a climate change journalist, puts it in a piece for *Vox*:

> *"Even given optimistic assumptions about decarbonization, we'll probably end up emitting a lot more than our carbon budget, so we'll need to bury between 100 and 200 gigatons of CO2 to get back within it. And, of course, we'll have to bury hundreds of gigatons more in the years after 2050. To give a sense of scale, that means by 2030 humanity needs to be compressing, transporting, and burying an amount of CO2, by volume, that is two to four times the amount of fluids that the global oil and gas industry deals with today. To build an industry of that scale, by that date, we need to begin today, with large-scale research and deployment. The price of capturing CO2 from the air needs to be driven down quickly."*[14]

Still, perhaps the most intuitive problem arising from the carbon capture agenda is (d), the fact that the process can only work to capture and remove levels of harmful carbon that already exist and are circulating within our ecosystems.

Maybe it's just me, but based on all of these respective factors, my immediate decarbonization question then becomes: Why not attack the problem at its origin, instead, while particulate matter and black carbon are still just unburned fuel, and prevent it from being released into the atmosphere at such a high volume in the first place? This can only happen if those fuel-burning engines and machines are able to burn their fuel better (meaning, more completely), which can only be accomplished by introducing a catalyst like hydrogen into the fuel combustion process (and voila, the solution).

For decades, hydrogen has been proven to improve fuel combustion and reduce harmful emissions in the labs of countless scientists and engineers across the globe. But only within those labs. There has never been anything commercially available that we could actually apply as a pollution solution to the huge spectrum and sizes of fossil-fuel-burning engines in the field. Until now. Until that day in the lab when a little fuel-saving device, now patented as the LeefH2, was found to reduce particulate matter and black carbon creation by almost 50 percent—and a simple, scalable, real-time solution to climate change was born.

That was when my focus turned from saving fuel and saving money to actively reducing and eliminating particulate matter and black carbon from the planet. And I have gladly never looked back.

CHAPTER TWO

Sorry Bernie, AOC & Others: Big Oil Isn't to Blame— It's Us (and YOU)

"We can blame the oil industry for a lot of things, but climate change is not one of them. Climate change is on all of us."—
Donald Owens

First of all, I am not at all positive that Bernie and AOC are blaming the oil industry for climate change, so I will apologize in advance if I have misinterpreted some of their politics, but I do know there have been some recent court decisions that have tried to make such a case. Hopefully, this chapter will dispel that myth and point out the real culprits of climate change, and those culprits can be seen in the mirror.

Now that we've refocused the discussion and identified the true, immediately addressable enemies of climate change (which are those nasty characters, particulate matter and black carbon)—it's time we also contextualize and correct the current narrative surrounding the fight against climate change, aka global warming. Many environmentalists would have us believe that the fossil fuel industry, or "Big Oil" as they so lovingly call it in America, is the dirty rotten step-uncle who nobody invited to the dinner party but who still proceeds to gobble up all of the appetizers, drain the liquor cabinet, clog the toilet, break a vase, make your granny cry and yet somehow still manage to win the state lottery on his ride home after you stuffed him into a cab. The industry has been painted with vivid, broad, scarlet brushstrokes to be the ultimate villain in earth's epic environmental saga. It has been made out to be the cannibalizing, capitalistic machine that ruins its own planet in a blind attempt to gain unimaginable power and profit.

"Big Oil" and the global fossil fuel industry may be to blame for many of the modern ills in the world (which I will not even begin to ferret out in this book), but it's not to be blamed for climate change. Climate change is on us. All of us.

Ever heard of the fundamental economic concept of "supply and demand?" In most societies, people create things that other people want / find value in, and the production of such goods is directly correlated to the demand for them.

Ask yourself this: Do you like having the simple freedom to choose whether or not to use a portion of your own earned income to heat your home when a blizzard is about to come your way or to give your children a warm enough bath at night to keep them from catching a cold? (Remember how, earlier, we teased to the uncomfortable but unavoidable truth that literally every modern comfort and convenience produces CO2 and/or black carbon pollution as a by-product of its development and/or consumption? Good. Hold on to that thought, because we are about to dive into it in much, much more detail very soon. So, hang tight).

All of these questions are somehow frequently forgotten, ignored or omitted from the current cultural conversation regarding climate change in this country, especially when highly influential people on both sides are too busy pointing fingers at each other to realize the simplest truth of the matter, which is this: ***No single human or country or industry is at fault for our failing environment. It is all of us.*** We are all at fault, all at once. Every single modern person who participates in civilized society to any degree is the direct cause of this harrowing environmental effect.

Allow me to explain. One of the most beautiful, differentiating aspects of a free market economy is the ability for independent organizations to identify a demand, to develop a good or service that can meet that demand to the best of their ability, and to then scale those goods or services for the audiences or consumers that required it in the first place (there is a reason it is called a "demand," and not a "request," after all). This is basic economics 101, and the microeconomics law of supply and demand forms the very foundation of America's great historical growth, grit, and global rank—year after year after year. Without our ability to have quickly scaled up our factories and manufacturing capabilities during World War II to meet the extraordinary demand for military equipment and domestic goods, for example, we likely could have lost our efforts (and far more soldiers) in that fight, and who knows where we would be today as a consequence. In fact, according to Doris Kearns Goodman, an American Pulitzer Prize-winning historian and social policy commentator, this quick industrialization and economic breakthrough was the very way in which we won the war in the first place. In a 2001 article she penned for *The American Prospect,* Kearns noted:

> *"America's response to World War II was the most extraordinary mobilization of an idle economy in the history of the world. During the war, 17 million new civilian jobs were created, industrial productivity increased by 96 percent, and corporate profits after taxes doubled. The government expenditures helped bring about the business recovery that had eluded the New Deal. War needs directly consumed over one-third of the output of industry, but the expanded productivity ensured a remarkable supply of consumer goods to the people as well. America was the only [nation] that saw an expansion of consumer goods despite wartime rationing. By 1944, as a result of wage increases and overtime pay, real weekly wages before taxes in manufacturing were 50 percent higher than in 1939. The war also created entire new technologies, industries, and associated human skills ... It is no exaggeration to say that America won the war abroad and the peace at home at the same time."*[15]

You see, America (and the rest of the civilized world) tends to do industry very well, and we mostly reap the benefits for it. But, as we all know, every upside has its equal opposite, and unfortunately, the downside to this competitive, continuous innovation can be very, very steep.

But can we blame the fossil fuel industry for the slick ride down that steep slope? Hardly. The fossil fuel industry is merely an explosive, long-lived example of successful supply and demand at play. If we still look just within the scope of American history and our domestic utilization of fossil fuels (which are petroleum, natural gas, and coal), we see a two-hundred-year-old survival story of innovators and tinkerers who learned to leverage nearby natural resources to meet immediate, basic human needs. For example, when William Hart of Chautauqua County, New York, first dug a deep well and borehole through his wife's washtub into gas-bearing shale in 1821 and discovered that he could use that natural gas as a source of illumination if he piped it up from the ground and burned it above the surface, he probably didn't realize he was initiating the American natural gas industry.[16]

He likely couldn't have predicted how quickly those around him would start digging additional wells to accomplish the same ends, thereby making a tourist attraction of the well-lit streets in the town he chose to drill in. He wasn't a politician or titan of industry or esteemed engineer at the time, either. Nope. Billy was just an imaginative gunsmith who saw an opportunity to extract flammable petroleum from the planet and use it to light one single inn nearby.

And when, almost thirty years later, Samuel Kier first discovered that a popular Kentucky-bottled curative was composed of the same dark, viscous substance that often mucked up the water in his family's brine wells and decided to capitalize on the coincidence by skimming the salted sludge out from those natural oil seeps to eventually repackage and sell as a similar medicine,[17] he probably didn't realize he would one day be lauded as the "Grandfather of the American Oil Industry."[18] Sure, he did start the first American crude oil refinery out of Pittsburgh, Pennsylvania, in the process, but his original intention in doing so was to create a surplus of the medicine that had eased his wife's pains from consumption (or tuberculosis), even without a background in science or medical chemistry—while making a pretty penny, of course.

Kier also probably didn't predict that the brine well wooden derricks he had used to pull the crude oil for his medicine would inspire two other Pennsylvania tinkerers, George Bissel and Edwin L. Drake, to later leverage a similar but more sophisticated apparatus to extract large quantities of rock oil for the specific purposes of distilling kerosene. Bissell, a lawyer, ended up forming what have since widely been considered the first two American oil companies, Pennsylvania Rock Oil Company and Seneca Oil Company, the latter of which employed Drake, a former train conductor, who pioneered multiple fundamental drilling technologies and techniques—including the first successful drilling rig used on the first American well dedicated solely to the commercial production of oil.[19] According to historians like Edgar Wesley Owen in his 1975 book, *Trek of the Oil Finders*, it is less important that Bissell and Drake's work led to the first commercial oil well in America, and more that it started a veritable gold rush of large-scale investment into commercial oil production and marketing. Owen put it more precisely when he wrote, "The importance of the Drake well was in the fact that it caused prompt additional drilling, thus establishing a supply of petroleum in sufficient quantity to support business enterprises of magnitude."[20] They crafted the commercial blueprint that was able to draw the likes of John D. Rockefeller to invest and strategically start scaling that Pennsylvania oil rush into the modern, global behemoth industry we know of pejoratively as "Big Oil" today—of which market research by IBISWorld estimated total global revenues to be about $3.3 trillion in 2019 alone.[21]

Are you starting to see a trend here?

These men didn't know of the deep, dark costs of the mass production and scaled refining of their crude oil discoveries when they took them to market. Of course they didn't. They couldn't have—not just because the long-term data wasn't yet available, but because not one of them was an environmental scientist or chemical engineer or respiratory health expert. They were small-town tinkerers who sought to remedy small-scale problems close to their own hearts and homes, more often than not, drawing from the by-product resources they found on their own properties and neighboring lands. While perhaps we can call them out for acting strategically for their own benefit with a profit in mind, we cannot credit them with malicious intent. Doing so would, candidly, be hypocritical of us—because, truly, aren't we all just looking for a way to pay our bills, provide for our families, and maybe make an impact in the long run along the way? Isn't that the very foundation of the American dream? We would all be guilty of the crimes we wish to accuse them of if we seek to blame the American oil and fossil fuel industry for our currently crumbling environment.

But we would also be remiss if we stopped simply at the starting point.

Much like the World War II-era American industrial boom we discussed at the start of this chapter, as our manufacturing abilities exponentially expanded, production increased and our industries evolved to provide for many modern comforts, so too did our collective carbon-coated footprint. And I'm not saying that in the "woo-woo" guilt-inducing way that we often hear that phrase tossed around today. I'm speaking to the facts alone: Over time, we reached a place where every single mechanism of production and consumption became tied to the creation, release, and accumulation of harmful black carbon in our atmosphere. Sometimes this product pollution was achieved indirectly, but more often, it has occurred immediately and visibly.

In 2021, we cannot stop the production of excess CO2 and black carbon, because almost every single thing we do on a daily basis creates excess CO2 and black carbon.

Think back to your day thus far. Did you wake up inside of a fully constructed residence, with walls and ceilings and doors and paint? (Factually, the construction and production of every part of your home released sizable amounts of CO2 and black carbon into the atmosphere.) Did you exit a mass-produced mattress model when you arose? (Or did you sleep on a pile of organic palm fronds you collected from the floor of a nearby jungle?) Does your bed have sheets, which you wash with any amount of regularity? Do you use detergent to clean them with? (Pro tip: Even if the detergent is "certified organic," "environmentally friendly," and "chemical-free," it doesn't preclude the machines and large vats that produced that eco-detergent from releasing harmful black carbon into the atmosphere since they are powered by non-solar electricity). Do you wash those sheets in a laundry machine (or do you use a washboard out back that you carved out of recycled materials)? How do you heat the water for the wash? And were you wearing any manufactured clothes when you did so? (Don't answer that.) Do you keep a phone charger or humidifier or alarm clock or Amazon Alexa plugged in within reach to get you through the night? If so, how did you first procure them? Did you have them delivered directly to your home? Or maybe you drove to the local superstore and picked them up off the shelves before purchasing. I wonder, how did those items make it to those shelves from their original manufacturer? (They weren't shipped on an electric, self-driving Tesla, I'll tell you that much.) Was your room heated or cooled to any degree for comfort, thereby making it all the more difficult to get out of the coziness of your bed? If so, how does the central heating or air conditioning or standalone device you have propped inside your window heat or cool the air? Does it use electricity from a solar panel on your roof? That's doubtful, but do you even know?

Of course not. We are already far too overwhelmed by minute details and endless streams of information to know the entire production chain of every single item we own or operate. Stop reading this book for a minute and take a look around. What do you see? If you are in an apartment or a house, you see furniture, from tables, desks, lamps, chairs, electronics like a television set, water bottles, alcohol, snacks or all sorts, pictures on the walls, clocks, plants, real or fake, trash cans, candles. The list of what you see goes on and on and is then multiplied by every other human being on this planet that does not live in the bush. The permutations for the infinite products and processes that surround each of us on a daily basis that required some carbon-creating mechanism to be involved in their manufacturing, packaging, shipping, transportation, stocking, selling, buying, and consumption are both endless and compounding. There is perhaps nothing more ubiquitous in our current society than the endless opportunities to unknowingly pollute as a modern civilized consumer. And we've only just highlighted a tiny portion of the process of simply getting out of bed in the morning. Pull that lens out a little wider to include our travel, commutes, commerce, holidays, and the other twenty-three hours of our everyday lives, and my goodness. We don't even have time for that kind of analysis. We would never be able to track it all.

But the truth is, we cannot go back.

Or, perhaps more accurately, we don't really want to.

As the wise Professor Milton Friedman so eloquently put it in his book, *There's No Such Thing as a Free Lunch*, "Even the most ardent environmentalist doesn't really want to stop pollution. If he thinks about it, and doesn't just talk about it, he wants to have the *right amount* of pollution. We can't really *afford* to eliminate it—not without abandoning all the benefits of technology that we not only enjoy but on which we depend." That was written in 1975. My God, what foresight.

Every innovation, every single miraculous feat of imagination that humankind has developed in the past century especially has relied on machinery, systems, and processes that produce harmful black carbon. Even those technologies and inventions that are ultimately aimed at reducing our carbon footprint and moving us collectively closer to a greener, 0 percent carbon emission future (what a lovely cotton candy dream!) contributed to the pollution that we are currently grappling with—because those factories that develop those electric vehicles (more on *that* later) still have to run on fossil fuels to some extent to power the production of their enormous equipment. The shipping vessels, trains, planes, and rigs that transport those eco-friendly parts and materials churn out CO2 and black carbon sludge like nobody's business. (It's also worth noting that these engines will likely never be able to fully switch over to electric energy, either, because their heavy loads require them to run on diesel fuel, which is one of the worst black carbon polluters.) Can you guess how much machine power, physical mass and weight capacity it takes to deliver a new wind turbine to its destination? The answer: it can take upward of 750–1,000 individual loads and trips (across rail, road and sea) to transport the massive pieces of one single windmill, with routes often originating from across the country or overseas—and I can tell you one thing for sure, they ain't using hybrids to do it. And what about the domestic recycling programs we so meticulously try to support and play our part in by sorting out our personal garbage into compost, trash, and recyclables bins? I hate to be the bearer of bad news, but those don't really do much of anything at all for the environment. According to Jonathan Clark's 2019 exposé article for *Medium:*

> *"Recycling programs are, with a few exceptions, largely a waste of time, money, and resources. In its current configuration, many programs are unsustainable sans large government subsidies and in many cases, a bit of a fraud ... The inconvenient truth is that, with few exceptions, mandatory recycling programs do little to help preserve the environment and in fact, many recycling processes may do more harm than good. And surprise! A growing portion of the trash deposited for recycling ends up in landfills."*[22]

But there's no way I can stop recycling. I like the planet too much not to and, fortunately or unfortunately, I like the idea that I am doing at least ***something***. If you're ready to bang your head against a wall at this point, you're not alone. I'm right there with you. But remember, the goal of this book is not to dwell in that helpless, frustrated feeling or point out some futility of existence. We are about hope and solutions and the future here. But we must correct the conversation and right-size the argument if we ever dream of getting there.

So, when a vocal environmentalist pickets a power plant or oil refinery or new pipeline site and shouts at us to boycott "Big Oil," they forget some important details:

- How did they, personally, get themselves to the site that day to picket?
- How did they procure the materials for their protest sign?
- How were those sign materials originally manufactured?
- Assuming they are not assembling alone, how did they all arrive at their protest spot? Did they caravan in a neat little line of fully electric vehicles only in the carpool lane? Did they charter a bus to try and be mindful of their commute's carbon footprint? If so, what are the chances that that bus or van or train has an electric engine? (Unless they are protesting in China, which boasts 99 percent of the world's electric buses, the answer is: highly unlikely.) Or did they drive themselves in their own good ol' gas-powered vehicle?
- What is their end goal? If they are successful in their call to boycott "Big Oil," are they comfortable living in a hut they sewed for themselves in clothes they must wash by hand inside a natural waterway and never showering or using an open flame to cook their meals or draw heat from ever again? Because fossil fuels power a lot of those alternatives, and an open flame *is* black carbon.

I'm sorry to break it to you, Bernie, AOC, and every single one of my beloved Democrats, Republicans, and Independents alike. ***We are the polluters***. You, and me, and everyone else. Our very way of modern life demands it. Please note: I did not say "requests" it. I said "demands" it. Our first-world dreams, desires, and devices demand it. In fact, virtually all of the technological and medical innovations that have helped extend our life expectancies and wildly decrease infant mortality rates depend on fossil fuel consumption. Any and every modern comfort can be traced back to a sizable contribution to climate change. And we like to be comfortable, don't we? We like to keep our loved ones warm, protected from the elements, fed and entertained? Yes, of course, we do. I was in Las Vegas recently. It was 115 degrees Fahrenheit outside. As I sat inside at a restaurant looking outside at the beautiful pool, I was very thankful that the 115-degree heat was not inside with me as I ate my homemade ice cream so wonderfully prepared by the restaurant's chef. If it had been, I would have been demanding that the hotel repair the AC and unfortunately I would not have cared if the electricity was from solar panels or coal. And I'm an environmentalist. But there I was, creating black carbon like all of the rest of the hotel's guests. Now that the pandemic is practically over, there were a lot of hotel guests in the city. Every last one of them (along with me) demanding the comforts that fossil fuel bring. Then multiply that by every city in America and every city in Europe and every city in South America and every city in Germany … and so on and so on. WE produce a lot of black carbon. As I sat there in the comfort of that hotel, I was glad someone was drilling for oil. So, it's time we stop blaming "Big Oil" for simply being in a business that supplies our ever-increasing demand for comfort and convenience.

That's not to say that massive organizations like those within the fossil fuel industry don't have a responsibility or a role to play in fighting climate change moving forward. I would expect those companies to heed these warnings and adopt our solution for eliminating black carbon creation (the LeefH2 device) just as much as I'd expect any individual or business that is made aware of these problems to take some responsibility and action. But we are all kidding ourselves if we think they (the oil industry) acted alone.

It was always us.

CHAPTER THREE

Pissing in the Wind: The Absolute Folly of Waiting on EVs

"We cannot solve our problems with the same thinking we used when we created them."—Albert Einstein

Before we dive into the good stuff re ramping up to green energy, it's about time to remind us of those two crucial, core statements about climate change:

1. ***Black carbon coats the glaciers, causing them to absorb light and heat instead of reflecting light and heat, and they melt. Period!***
2. ***Any climate change "remedy" that does not directly address black carbon is pissing in the wind (wind from a Category 6 hurricane, in fact, with the wind from that hurricane blowing directly in our face).***

By now, you might be thinking, "Well, Don, fine, the climate is collapsing, and we are all at fault, so I can't just keep blaming big bad Big Oil. OK, fair point. I'm with you. But what about the imminent rise of electric vehicle adoption? That's a really hot topic right now, right? Isn't that the answer? Won't we all be driving fossil fuel-less cars soon enough, anyway? I thought I'd heard that that would be the best way for me to get involved and play my part in solving for harmful CO2 emissions and black carbon creation in the immediate future?"

I'm glad you asked. Because there's nothing "immediate" about that game plan.

Sure, the California governor did just pass a landmark executive order in 2020 requiring that all new cars and passenger trucks sold within the state are to be zero-emission vehicles by the year 2035, for example. I can't fault the guy for his proactive and wishful thinking, but let's be real for a minute.[23]

The automotive industry typically sells about 16 million vehicles annually in the United States alone, with California accounting for about 2 million of those sales (even in 2020, in the midst of the pandemic, when most travel had totally halted). That same year, California, leading the charge toward green energy, sold about 240,000 electric vehicles (EVs), plug-ins and hybrid vehicles, according to the California New Car Dealers Association—which only accounts for about 12 percent of those overall car purchases. Assuming that these trends remain fairly stable, it is relatively safe to assume that in the fourteen years remaining between now and the 2035 goal date, Californians will purchase upward of 30 million vehicles, while the rest of the country will purchase somewhere around 224 million vehicles. Considering that each of those freshly purchased vehicles carries an expected lifespan of at least fifteen to twenty years, almost all of them will still be on the roads by 2035. Even if we bump that current 12 percent EV rate up to, say, a hefty 50 percent, it would still require some degree of pure, unadulterated fantasy to surmise that those many millions of nonelectric vehicles (which all consume either gasoline or diesel fuel) are just going to somehow disappear. And that fun little thought experiment doesn't even address any of the other parts of the United States or the rest of the world. ***That only reflects California. For the rest of the US, the EV rate is more like 2 percent. And that does not include the rest of the world.***

But, sure, let's go ahead and fool ourselves for a few minutes into wrongfully thinking that switching to EVs is something to reasonably consider. I won't even mention the fact that the electricity to charge the batteries of the electric vehicles does not come from "renewable" sources.

Those batteries are being charged by the US electrical grid, where around 20 percent comes from renewable sources. So the black carbon has shifted from the tailpipe to the power plant. But it's still being generated (but I'm not even taking that into account below). Please understand, I am NOT in opposition to electric vehicles. But trying to combat climate change exclusively with today's renewable energy technologies simply won't work; we need a fundamentally different approach.

Unless the government starts seriously subsidizing public income levels and/or the average cost of buying a brand new vehicle (while simultaneously somehow easing the sentimental value of family heirloom models and classic car nostalgia), there is no way they will be able to limit the necessary sales and purchases of existing fossil-fuel-powered cars or trucks. (Unless I suppose, they were to negatively incentivize folks by ticketing every gas engine vehicle they find on the road, but that would lead to a political and social upheaval that is beyond the scope of imagination.) Think about it. Most average income earners cannot currently pay cash for a new vehicle, and the youngest consumer generation seems to be much warier of accumulating routine debt. We are even seeing a large swath of Millennials and Gen Zers forfeiting the traditional American lifestyles of nuclear families, picket-fenced home mortgages and nine-to-five corporate careers in favor of remodeling old school buses as tiny homes and traveling along the backroads of the country with their dog in their lap while building out an app or social media following from their phone or laptop whenever they can siphon off someone else's Wi-Fi connection. Yes, they also tend to be more environmentally conscious, but last I checked, those buses and vans still run on fuel.

Think back to your own first car, even. Was it a luxury leased vehicle fresh off of the lot? A shiny new pickup with custom wheels and a hefty down payment? If so, I applaud you, for that's quite a life you've lived. But, if not—if you're like me and most of the general population—your first car was likely an older but reliable model that cost somewhere between $5,000 and $10,000 and could be bought used off of the "classifieds" section or from your neighbor down the street. Why? Because that's what we could afford at the time, and it was all we needed to get us from point A to point B. So, I wonder, do these eco-vocal politicians just expect that, suddenly, people will no longer choose to purchase older models that predate their newest environmentally-minded mandate? And even if they could get unanimous, enthusiastic consent and buy-in from the entire general population, do they expect that we could successfully make that massive transition in production, distribution, and maintenance in a little over one decade's time? (We would, after all, need to replace about twenty-eight million vehicles in California alone, which is not going to happen.) AND, assuming we could, in fact, kick our electric vehicle production into overdrive to match pace with that aggressive timeline, do they not realize the sheer amount of CO2 and black carbon that will be released as a direct consequence of that rapid ramp-up production effort? (Again, per our previous analysis of the subject, those factories are not exactly "solar-powered" or independent of fossil fuel consumption right now).

Which brings us to the true crux of the issue and the title of this chapter: we are foolish if we think waiting on green energy and/or electric vehicles alone will solve our enormous climate crisis in time for us to truly continue living on this planet—for a myriad of reasons, not the least of which is the fact that these consumer vehicles contribute only a small fraction of the total black carbon creation and excess CO2 emissions globally (including the other elephants in the room, which are construction, shipping, mining, power generation, agriculture, etc.), and the amount of pollution that would be released from a rapid, radical adoption scenario like this (which fails to address black carbon) could likely kill our chances of earth-bound survival sooner than later. Let us explore.

First on the docket of our analysis is this question of how much our commuter vehicles (or light-duty vehicles) actually contribute to the overall global warming trend in America and worldwide each year. While many reports find averages for the transport sector's total contribution to climate change falling within the 22–28 percent range (roughly one-quarter of civilization's overall warming impact on the climate), commuter vehicles only represent roughly half of this percentage. Furthermore, despite the fact that heavy-duty vehicles (big rigs, buses, trains) only account for about 5 percent of all vehicles on the road today, they can generate a substantial chunk of the overall harmful greenhouse gas and global-warming-causing emissions that come from that entire sector, including a significant amount of the air pollution that springs from global transport. According to a 2018 report published by the Union of Concerned Scientists:

> *"As the United States moves more and more freight each year, the challenge of reducing emissions from this sector will continue to grow. Addressing heavy-duty vehicle pollution is critical for improving air quality and reducing global warming emissions in communities around the country."*[24]

So, if we would like to continue ordering our Amazon Prime packages and subscribing to HelloFresh meal delivery services, we must also be increasingly mindful of the way such modern buying habits are impacting the increase in heavy-duty vehicles on the road to support that demand, along with damaging effect those vehicles have on our planet currently.

Now, some efforts have already been made to improve the fuel economy, decrease the particulate matter exhaust and update the internal pollution controls for newer models of these heavy-duty emitters, but unfortunately, these efforts have been neither sufficient nor sustainable—especially in light of how many of those fixes have failed since being implemented, along with how weak (or nonexistent) our monitoring and maintenance of such measures currently are. According to an article on *Inside Climate News*:

> *"Pollution controls that were built into heavy duty trucks nationwide beginning in 2007 began failing on some of those trucks within a decade, leading to significant increases in emissions of*

black carbon from some of the vehicles, a new study has found. The results suggest that tens of thousands of trucks sold in the United States between 2007 and 2009 could be operating with failed emission controls and are likely to remain on the road for decades ... The study found that the controls worked initially, but that their average black carbon emission rates jumped by as much as 67 percent between 2013 and 2015. High-emitting trucks made up just 7 percent of the fleet at the port, but they were responsible for 65 percent of the fleet's black carbon emissions."[25]

An air quality manager with the Environmental Defense Fund even pointed out, "If we are starting to see that those emission control technologies don't have the same service life as the engines themselves, then we are working backward in terms of the outcomes that we are looking for with respect to air quality." Add to this the fact that there are practically no emission abatement procedures currently in place for these trucks (and those other major elephants in the room) and, Houston, we have a problem.

One group of atmospheric scientists and engineers in Europe even went so far as to attach a makeshift bouquet of air intake tubes to the window of a passenger minivan to test and measure how much particulate matter was truly being released by more recent models of heavy, diesel-fueled trucks and buses under regular road conditions. Their findings were a bit unsettling, as they discovered that the efforts by manufacturers to decrease other pollutants may have actually accelerated black carbon production as a by-product. According to their study:

"Newer, diesel-fueled, heavy trucks and buses emit, on average, 34% more of the health and climate hazard known as black carbon than older vehicles of the same types. The unexpectedly dirty exhaust from heavy vehicles newer than 5 years old, compared with that from 5- to 10-year-old vehicles, may indicate that modifications by vehicle manufacturers to lessen other pollutants have

> *had the undesirable side effect of boosting engines' black carbon output, the researchers suggest."*[26]

It is also important to note that all of these factors, projections and estimates above are addressing the total muddied cocktail of environmental elements that intermingle and mix together to contribute to climate change, meaning: excess carbon dioxide, hydrocarbons, nitrogen oxides, and the "super pollutant" particulate matter (i.e., black carbon). Remember how, earlier, we outlined that black carbon is actually the worst of these ingredients, as far as human health and heating capacity to melt the glaciers is concerned? Well, when we look more specifically at pollution in terms of black carbon production, we find that the total consumer transport sector only emits about 19 percent of global black carbon annually.[27] Which means that there are much bigger fish to fry (81% bigger) than consumer vehicles in slowing climate change. Urban bus fleets, for example, are estimated to contribute a whopping 25 percent of the entire transport sector's black carbon content. According to the Climate & Clean Air Coalition, "Due to rapidly growing urban populations and increasing demand for efficient and affordable mobility, urban bus activity is predicted to increase by nearly 50 percent by 2030. This will translate into an estimated additional 26,000 tonnes of black carbon emitted in 2030."[28] And here, we thought we were decreasing our carbon footprint by opting for community transport over individual vehicles. How silly (but want-to-do-the-right-thing) of us.

Second, we can look to that earlier question of how realistic it is to transition to full-scale, mandated 100 percent zero-emission commuter vehicles by the year 2035, as California is aiming for. According to Ian Simm, CEO of Impax Asset Management (which is responsible for managing something like $18.5 billion assets, including substantial investment into water and natural resources), as quoted by the Dow Jones legacy market and investment publication, *Barron's*: "While transportation technology is just about [to the place to enable widespread electric vehicle availability], it will take until the 2040s ... to get costs down, so that electric vehicles are as cheap as internal-combustion cars, buses, and trucks. There's also the legacy-asset issue; just about every American drives a gasoline-powered vehicle now, many of which will last another 10 or 20 years."[29] (We called that last part, didn't we?!)

But that's just as far as the consumer's budget is concerned. In order to truly reach an effective rate of adoption for green energy initiatives to meet these idealistic expectations, we need significant participation and partnership from the entire industrial sector—and let's be real, buy-in for a massive overhaul of any system, structure, or process in business requires substantial, measurable proof of ROI and impressive financial or tax incentives to gain any traction in the boardroom. In fact, two of the engineers who worked on one of Google's biggest energy initiatives, RE<C, shared exactly why motivating businesses' bottom line is a huge piece of the climate change puzzle in their reflections on the failed project for IEEE Spectrum when they said:

> *"Even if every renewable energy technology advanced as quickly as imagined and they were all applied globally, atmospheric CO2 levels wouldn't just remain above 350 ppm; they would continue to rise exponentially due to continued fossil fuel use … What's needed, we concluded, are reliable zero-carbon energy sources so cheap that the operators of power plants and industrial facilities alike have an economic rationale for switching over soon—say, within the next 40 years. Let's face it, businesses won't make sacrifices and pay more for clean energy based on altruism alone. Instead, we need solutions that appeal to their profit motives. Consider an average US coal or natural gas plant that has been in service for decades; its cost of electricity generation is about 4 to 6 US cents per kilowatt-hour. Now imagine what it would take for the utility company that owns that plant to decide to shutter it and build a replacement plant using a zero-carbon energy source. The owner would have to factor in the capital investment for construction and continued costs of operation and maintenance—and still make a*

profit while generating electricity for less than $0.04/kWh to $0.06/kWh. Similarly, we need competitive energy sources to power industrial facilities, such as fertilizer plants and cement manufacturers. A cement company simply won't try some new technology to heat its kilns unless it's going to save money and boost profits. Across the board, we need solutions that don't require subsidies or government regulations that penalize fossil fuel usage. Of course, anything that makes fossil fuels more expensive, whether it's pollution limits or an outright tax on carbon emissions, helps competing energy technologies locally. But industry can simply move manufacturing (and emissions) somewhere else. So rather than depend on politicians' high ideals to drive change, it's a safer bet to rely on businesses' self interest: in other words, the bottom line."[30]

Last, and perhaps most crucial, on our list of reasons why it is foolish to wait on widespread adoption of electric vehicles and green energy is the reality that the rapid production necessary to meet those timelines will, in a great twist of irony, accelerate the rate of global warming dramatically. Take solar and wind energy, for example. Now, as great as these technologies are, they really only work under certain circumstances to begin with. A windmill can only turn when wind is present (and can only be equally powerful to fuel-generated electricity in especially windy places), and a solar panel can only collect sunlight when the sun is actually shining. This already leaves a lot of energetic downtime in the equation. But have you ever actually considered the sheer size and magnitude of even a single wind turbine? How do you imagine those massive materials are manufactured, transported, and assembled? If you think we can build these behemoth undertakings in the absence of fossil fuels—gosh, I really hate to be the bearer of bad news again. But I don't have to break it to you myself, because the smart guys at the Barnett Shale Energy Education Council already did it for me:

> *"The reality is that wind and solar are dependent on fossil fuels. They cannot exist without oil and natural gas ... Wind turbines and solar panels cannot be made solely from other wind turbines and solar panels. Wind and solar facilities currently require massive quantities of steel and concrete, both of which require oil and natural gas in their manufacturing processes. The amount of steel required for wind and solar to replace fossil fuels exceeds the world's capability to produce it for decades. An article in Forbes said 'the unavailability of sufficient steel prevents wind from replacing just coal, since that would take 10 billion tons of steel. The total annual global production of steel is only 1.6 billion tons.' State-of-the-art wind turbine blades are made of carbon fiber, which consists of layers of plastics and plastic resin, both of which are derived from oil and natural gas feedstocks. [Furthermore], many components of wind turbines and solar panels are manufactured in China and transported to the US in ships that burn heavy fuel oil or diesel."*[31]

This moves our attention from the reliance on fossil fuels that the very production of these green energy solutions require to the heavy diesel fuel usage and consequent black carbon creation that stems from their necessary transport. According to a 2017 report released by Lockheed Martin, "the global market for wind energy continues to grow in excess of 10 percent per year. This means as many as 10,000 new turbines will need to be installed around the world over the next 20 years."[32] When you then pepper in the truth of how many fossil-fuel-powered fleets it takes to get those wind turbines from their origin to their destination, it is hard to deny that this process doesn't quite jive with the "carbon-neutral" and "carbon-negative" narratives its green solutions typically espouse. According to that same report from Lockheed Martin, "Currently, wind components are transported using a variety of different modes, including ship, rail and truck. For example, a 150-megawatt wind farm can require as many as 650 truckloads, 140 railcars and eight ships to complete the transportation process." That is a whole lot of net-new black carbon on our roadways and in our oceans.

What can be said after learning (and actually understanding) stark realities like this? The phrase "HOLY CRAP!" comes to mind. I think we can all agree that climate change is indeed a huge, dangerous, and unruly monster that we have to try our best to tame as soon as possible. But Bernie, AOC, Greta, and every last one of us who just wants to do the right thing about climate change (because I know all of you do) must realize: ***if we don't address black carbon, we are indeed "pissing in the wind."***

So, in conclusion, does waiting to convert all consumer vehicles to electric—a conversion that cannot possibly keep up with the rate of the worldwide consumption of gasoline and diesel consumer vehicles and which represents such a small fraction of the black-carbon-producing world of vehicles and machinery that make life livable on earth and which actually causes more black carbon to be created in ramp-up in the making of these electric vehicles—make sense? As candidate Joe Biden once said before becoming president (and as many of us have said): ***"Come on, man!"***

CHAPTER FOUR

Hydrogen: Its Glorious Future and Its Inconsequential Present

"I strongly believe that the next step of the global energy transition will be based on the hydrogen economy ... via water electrolysis. These chemicals can be transformed, stored, transported and used in various sectors. This sector coupling approach allows us to decarbonize applications whose electrification will come to its limits." —Prof. Dr. Armin Schnettler, EVP & CEO New Energy Business, Siemens Energy

The hydrogen economy is an envisioned future in which hydrogen is used as a fuel for heat and hydrogen vehicles, for energy storage, and for long-distance transport of energy. Water electrolysis (the conversion of water to hydrogen) is a technology that can extract hydrogen from water and use it for some of the stated uses above. As of today, however, there are NO major hydrogen installations or processes that would slow down our rapid unobstructed march toward global warming and climate change. This is about to change because of the development of the LeefH2 (discussed in the next chapter), but as of the printing of this book, the hydrogen economy is a welcomed vision of a clean hydrogen future but it will take decades, if not centuries, before it can have a positive effect on our environment or climate change. However, it is still very much worth talking about.

Hydrogen is the lightest, most common, and most abundant element in the known universe. The colorless, odorless, unassuming stuff is so ubiquitous to our existence on earth (and beyond) that almost all of our chemical fuels spring from and are bonded with hydrogen. Nature is essentially tripping over itself with a bounty of hydrogen. Hydrogen also happens to burn faster, cleaner, and more completely than any other fuel on this planet by orders of magnitude. It's magnificent. However, hydrogen does not exist or manifest on its own anywhere organically, so it must be isolated and extracted by man-made processes in order to be directly leveraged in its solitary form—which (in the past) required a good deal of extra energy to accomplish. As scalable systems for hydrogen extraction and application have begun to take shape and gain momentum over the past couple of decades, we've seen a demonstrated rise of global interest in its potential. We've heard a lot of buzz and excitement around a "clean hydrogen economy" and the endless possibilities that a hydrogen-powered future could promise the planet and the infinite life forms it is home to.

As a future fuel, there are truly no limits for hydrogen. In the present, however, there has been very little for hydrogen to do. As of May 2021, there were fifteen thousand plus hydrogen fuel cell vehicles on roads worldwide, with existing government targets aiming for that to increase to ten million by 2030. Not even a drop in the bucket in regard to impacting climate change. Even the ten million hoped for by 2030 is still not a drop in the bucket. Since the late 1990s, there has been huge investment in this technology by carmakers, but only Honda, Hyundai, and Toyota make fuel cell cars that you can buy today. Large-scale introduction has been hampered not so much by the tech but by the difficulty and cost of establishing a filling station network.

Up until recently, though, the only truly commercialized application of hydrogen's fuel optimization capabilities has been through the advent of hydrogen fuel cell technology, which turns hydrogen into a fossil fuel alternative by fusing oxygen and hydrogen together and converting the chemical potential energy stored within their molecular bonds into electrical energy. The capacity for storing and optimizing hydrogen within these fuel cells is impressive, as the fuel cells "pack more power into the same amount of space than electric batteries ... supposedly making hydrogen better suited for airplanes or ships that have to carry energy supplies long distances."[33]

Some of the clearest benefits of hydrogen fuel cell technology include the extreme availability of hydrogen itself, the renewability of energy produced in the process, the massive reduction in fuel consumption, and the minimal-to-zero carbon footprint and emissions created as a by-product. One report highlighted the faster charging times of fuel cell power units over battery-powered electric vehicles, stating that its extreme charging rapidity matched only that of conventional internal combustion engine (ICE) vehicles.[34] According to the report, "Where electric vehicles require between thirty minutes and several hours to charge, hydrogen fuel cells can be recharged in under five minutes. This fast charging time means that hydrogen powered vehicles provide the same flexibility as conventional cars." Beyond the superior charging capabilities of these systems, the report also outlined the longer usage gains from the optimized technology by stating that "a hydrogen vehicle has the same range as those that use fossil fuels (around 300 miles) ... [which] is superior to that currently offered by electric vehicles." In fact, it went on to describe how EV developers have actually been leveraging fuel cell power units to extend the driving range beyond the manufactured mileage capacity. That sounds promising, doesn't it?

A few major motor companies thought so, as they began rolling out their own iterations of hydrogen-powered internal combustion technologies over the past fifteen years or so. The great initial hope was that these hydrogen-centric systems could essentially power vehicles in the same way that gasoline could on stores of the periodic table's first element.

BMW, for example, introduced a version of the 2005–2007 BMW 7 series, the Hydrogen 7, with a 6.0-liter V-12 engine that could run on either gasoline or hydrogen. It made bold claims of about 40 percent efficiency when fueled by hydrogen versus the considerably lower fuel efficiencies seen in most gasoline engines. Mazda was particularly active in this arena and once claimed that its Wankel rotary was especially well suited for hydrogen operations (as the engine design already tended to run cooler than conventional piston engines, and could therefore reduce concerns over NOx emissions). The company first released a bifold version of its RX-8 that could transition from gasoline to hydrogen (and vice versa) when needed, then again in its Premacy H2 RE model as part of a series hybrid system. The ultimately flatlining factor for each of these vehicles—separate from their sheer cost—was the impracticality of storing enough hydrogen reserves within the vehicles to see meaningful results in range. After all, the Hydrogen 7 could only drive 125 miles on 17.6 pounds of hydrogen, after which gasoline would need to kick in to continue the journey.

More recently, the efficiency of hydrogen fuel cell vehicles has passed the 50 percent mark, meaning that more than half of the energy contained within the hydrogen is employed to operate the vehicle. In comparison with the Hydrogen 7's fuel economy, for example, today's 2019 Hyundai Nexo fuel cell vehicle can go up to 380 miles on 13.7 pounds of hydrogen. And Hyundai has recently said that the fuel cell stack itself in the Nexo runs at up to 60 percent efficiency. We're indeed getting closer in our efforts, yet still—there were only about seven thousand of these hydrogen fuel cell vehicles on the road in the US before the pandemic, and storing hydrogen within a consumer vehicle still remains an important obstacle that must be overcome to scale.

Each of those efforts, improvements and grandiose ideas are all well and good, but they still leave two almost unspeakably glaring gaps behind them (aside from pissing in the wind), both practically and scalably speaking: (1) sizable infrastructure issues and (2) hard-to-abate heavy emitting industry sectors, which contribute about 40 percent of global greenhouse gas emissions and release a substantial amount of black carbon behind them in their wake. That last detail is particularly troubling since currently, there is very little information, data, or research that's publicly available about exactly how much black carbon these hard-to-abate sectors create and contribute overall. If this doesn't change soon, we'll lose the battle before we've even really begun fighting.

To better contextualize and further evidence that statement within the context of hydrogen fuel cells, let's look a bit more closely at those two present gaps that stem from the currently available fuel cell technologies.

1. **Sizable Infrastructure Issues**

While governments, legislators, lobbyists, and private companies have been collectively pushing for a rapid ramp-up effort to install more EV charging stations across the globe in recent years, we have not seen a parallel rise in refueling locations for vehicles that run on hydrogen fuel cells or hydrogen internal combustion engines. Assuming basic consumer adoption, behavior, and use that would require a quick top-off or recharge here and there in city, suburban, or rural areas that may or may not yet be outfitted for such systems, drivers could be left out of luck if they wish to take even a brief road trip in their carbon-friendly "green" vehicles. This consumer need would need to be met in a massive way before hydrogen fuel cell vehicles will be able to offer a realistic and scalable solution for cutting down carbon footprints, but could take years (or even decades) to plan for—let alone actualize.

Also, while we're looking at the mass implementation of EV charging stations worldwide, it's worth noting that our existing gasoline and diesel refueling infrastructure is vast, intricate, and highly profitable. Any solution that cannot either retrofit these available structures and/or optimize existing fuel engines would be an egregious waste of both resources and opportunity.

2. **Hard-to-Abate Heavy Emitting Industry Sectors**

Despite all of our past and present advancements, the heaviest emitting sectors (mobility, mining, manufacturing, construction, freight, shipping, power, industrial processing, etc.) still remain largely out of reach/scope of any current environmental solutions, due in part to available technologies (or lack thereof) and the sheer cost of implementing these sustainable changes. Even hydrogen fuel cells have not yet been able to scale up in such a way to address these heavy polluters—these legacy machines and industries that we *need*, and which aren't going anywhere anytime soon. Left unabated, their substantial greenhouse gas and black carbon pollution will continue to cast a dark shadow over any other individualized improvements we can make in the fight against global warming.

To illustrate the magnitude impact and environmental implications of these hard-to-abate sectors, allow me to give you one simple use case: shipping containers. These gargantuan cargo ships that cross our oceans to bring us the goods we crave are essential to our economies, industries, and general way of living, but they are almost unparalleled in their pollutive capabilities. According to The Essential Daily Briefing:

> *"Every day, the clothes, tech and toys that fill the shelves in our shopping centres seem to arrive there by magic. In fact, about nine out of 10 items are shipped halfway around the world on board some of the biggest and dirtiest machines on the planet.* ***It has been estimated that just one of these container ships, the length of around six football pitches, can produce the same amount of pollution as 50 million cars. The emissions from 15 of these mega-ships match those from all the cars in the world.*** *And if the shipping industry were a country, it would be ranked between Germany and Japan as the sixth-largest contributor to global emissions. Most of the pollution occurs far out at sea, out of the sight and minds of consumers—and out of the reach of any government."*[35]

Needless to say, if we aren't addressing implementable change for just these cargo ships today, it doesn't really matter how much we can scale up or subsidize EVs and renewables at the consumer travel level. The black carbon these beasts release at sea is the exact same stuff we are seeing litter our polar ice caps and cause them to melt much more rapidly than was ever naturally intended. These form just one of the many pieces of the hard-to-abate sector puzzle, and yet, they are largely left out of the discussion. How unfortunate.

A duo of reporters from *GreenBiz* shared this perspective when they wrote:

> *"Hard-to-abate industry sectors constitute 40 percent of global greenhouse gas emissions. These sectors need a comprehensive approach to decarbonization, and they need it now. They need electrification, but they also need clean molecular energy. These are not mutually exclusive efforts, but can, when structured correctly, work in coordination to facilitate the rapid change that is called for. The pathway to scaling up green hydrogen will require substantial buildout of our renewable energy resources across the globe and the understanding that there are still battles to be fought to meet these targets."*[36]

Listen, I don't need you to take my word as gospel—but I'm also not alone in these estimations. The harsh truth is that this problem is far, far greater than consumer vehicles or most of the legislative proposals currently on the table, and the time to take decisive action in the right direction (that is, toward eliminating black carbon) is running out quickly.

The other, less promoted side of hydrogen, the uncapped power and potential of leveraging hydrogen as a means of cleaner fossil fuel optimization, is no secret. It has been well known and documented within the scientific and engineering communities for decades. Hydrogen has always made fuel burn better, but it has never been commercialized (until now).

It has served as the catalyst for countless discussions and debates surrounding diesel fuel combustion efficiency, along with every other hydrocarbon-based fuel. Historically, though, hydrogen as a scalable, sustainable environmental solution has been largely dismissed or discarded as an unrealistic pipe dream by many world leaders, government committees, and public officials, that have viewed any hydrogen-centric action plan as either too expensive or too inefficient to actualize in any meaningful way. And historically, they've had a point.

That is because traditionally, the production of hydrogen via extraction from fossil fuels has required greater volumes of energy than it has released upon consumption—thereby yielding net-zero or net-negative energy gains when all is said and done. These processes have also typically created excess CO2 and other greenhouse gases as waste by-products, effectively making it a contributor to pollution instead of an environmental solution. Because hydrogen at its normal temperature takes up so much physical space, also, the compression, liquification, packaging, storage, and transportation of any quantity of it has usually been a logistical nightmare. Lastly, considering that most past processes for hydrogen extraction have been so prohibitive in cost, it has routinely been cast aside and cut out of the conversation before it's ever really had the chance to prove the long-term value of any initial investments. For years, critics and experts alike have said hydrogen simply isn't worth the effort.

And until now, it hasn't been worth the effort. But now, in comes the first real scalable "hydrogen on demand" system that can extract hydrogen from water at very low energy levels and can be adapted to optimize the burning of fossil fuels in any engine on the planet. Due to its extreme flammability, when hydrogen is added to classic fuel combustion systems, it can optimize the burning of those fossil fuels so cleanly and completely that it significantly reduces greenhouse gases and particulate matter pollution by preventing a great deal of that unburned fuel (back carbon) from ever being created in the first place—even in "fuel lean" conditions, where there's more oxygen present than fuel.

Now hydrogen finally has something to do, which is saving the planet.

As much as EV advocates may hate to admit it, internal combustion engines are here to stay for the foreseeable future—and their applications reach far beyond gas- or diesel-powered vehicles. While the world is working together toward a cleaner-energy, zero-emissions future by the year 2050, we simply cannot get there by leveraging just wind and solar energy renewables. We also cannot expect countries to forfeit existing infrastructures and ecosystems made up of billions upon billions of dollars' worth of legacy assets in order to achieve that goal. In light of those details, hydrogen is undoubtedly the fuel of the future, but if we do the work today to scale up existing hydrogen generation solutions (like the LeefH2, which is explained in the next chapter) that optimize fossil fuel combustion across sectors and directly cut out black carbon and particulate matter at the source by burning fuel better, we can capture the entire addressable market and give hydrogen a supercharged, life-saving role to play here in the present.

Hydrogen, in each of these forms and systems, is the key to decarbonization at scale, but bringing its potential from the future into the present day requires every single player to agree and adopt all of these available solutions right now. Ultimately, the fight against climate change is not an either/or scenario. Renewable energy and cleaner fossil fuel optimization are not mutually exclusive avenues for our forward march. In fact, our only hope of overcoming the challenges of each is to employ both tools in tandem. There is no fork in the road where these paths must diverge from one another. This is not a competition, either. It is a singular, comprehensive, collaborative charge that requires all hands on deck.

That is simply the only plan of action that has any chance of securing a future for humanity on earth.

The plan has to include burning traditional fossil fuels better (through hydrogen electrolysis versus fuel cells alone) that could consequently cut out or cut down the creation of black carbon and CO2 during combustion. We could capture 100 percent of the opportunity at hand, which includes building out net-new renewable refueling and recharging stations while maintaining the operations and availability of existing gas station infrastructures. In this case, hydrogen would be given a present purpose, not just a future hope, and everybody wins—especially Mother Nature.

To paraphrase Gandhi, "If I want to see a change in the world, I have to be willing to start it myself." Which means I'll continue repeating, reiterating and emphasizing the immense importance of identifying and tackling black carbon as the true enemy of global warming and climate change until people actually begin to take action to address it. Because (as we have stated so vigorously in the previous chapter) any environmental action plan to combat climate change that does not directly address the concentrated power of black carbon in warming our planet is merely an exercise in pissing in the wind and will never meaningfully remedy our weakening environment on its own—at least not in time to achieve a positive, lasting, life-saving effect.

CHAPTER FIVE

LeefH2: The Solution to Climate Change is Here

"There is only one thing stronger than all the armies of the world, and that is an idea whose time has come."—Victor Hugo

Helplessness to hopefulness. Feelings of futility transformed into inspiration for action. Conceptualizing and contextualizing the true issue and enemy of the planet and the people in relation to climate change.

We've been through the gamut already in the course of this book, and there is a very intentional reason I've been repeating certain statements throughout the process. I've reiterated the fact that black carbon, which is created by literally everything that burns fuel, is falling onto and coating our glaciers in a way that is causing them to melt (and lining our lungs in a way that causes upward of ten thousand people to die every single day).[37]

I've also reemphasized my desire to relay all of these bleak, uncomfortable truths about global warming and the discussions surrounding it in such a way that could bring hope to those who have previously felt helpless about our planet's climate crisis. That is because, up until the latter portion of the last decade or so, I too felt pretty hopeless, helpless, and in the dark when it came to the issue of our rapidly warming atmosphere. Like many of you, I saw global warming as this big, Herculean beast that absolutely dwarfed me in its hot shadow—and most typically feel better if I left the matter ignored or unspoken of instead of ruminating on how futile any effort I could make to remedy it would feel, thinking at the time my only recourse was to recycle my plastic bottles.

For incredibly complex problems, such as our current environmental crisis, we often assume a similarly complex solution is required. And if we are not scientists, politicians, titans of industry, or philanthropists with hefty charitable foundation sums to invest at our disposal, it is easy to feel like the matter is beyond the scope of our ability to improve in any demonstrable, lasting, change-making way. But that's just the thing. We've discovered that the solution to climate change is actually much, much simpler than we could have ever expected it to be, and we have already actualized its production. **READ THAT AGAIN**.

Right now, here, today, we have a device available that is capable of making black carbon emissions a thing of the past by leveraging one of the earth's most abundant resources, water, while creating and releasing a surplus of oxygen for the planet in the process. That solution, our LeefH2, is real—it's ready to take up the fight to snuff out black carbon right at the source, and it is currently the ***only*** commercialized, customizable, hydrogen-based fuel augmenting solution that addresses this key black carbon component of climate change in an actionable, scalable way.[38] If implemented and mass adopted in real time for any and every fuel-burning engine globally (targeting larger diesel fuel engines first), we could actually see a shift in the harsh projections about climate change. We could see a marked improvement in fuel economy, cost savings, and decreased harmful pollution production across the globe. We could gain back time to reasonably ramp up our best green energy efforts. We would see clearer skies, colder (and cleaner) ice caps, more robust vegetation, and healthier lungs worldwide. Does this sound too good to be true? I'd probably have thought so, too, not too long ago. But it is good, ***and*** it is true.

The LeefH2 (meaning "Leveraging Energy Efficient Fuel with Hydrogen") device is a hydrogen generation/delivery system. Hydrogen has been proven repeatedly in university and engineering labs across the world to improve fuel engine combustion and performance while dramatically reducing overall emissions. Specifically, it has been proven to cut particulate matter and black carbon creation off right at the source, before it ever has a chance to disperse into our breathable air or be (mostly) filtered out by existing carbon filters, by up to 50 percent. And it accomplishes this simply by burning fuel in existing fuel engines better, faster, and more completely.

While other "clean diesel" engine technology attempts to catch emissions, particulate matter, and black carbon residue with filters in the engine's exhaust pipe, this still allows all the nasty stuff to be created in the first place. While these filters have been fit to diesel engines for about twenty years now, there are several maintenance and efficacy downsides to their use (some are rather serious for the lifecycle of the vehicle or engine), including routine clogging, susceptibility to tampering, ineffective self-maintenance, and zero mandatory performance checks in the United States. Often, a clogged filter can actually increase exhaust emissions and even occasionally throw the vehicle into what is called a "limp-home mode." Perhaps ironically, the travel conditions that typically develop the most blockage and disrepair to diesel particulate filters are those that are both short and kept at low speeds. A drive down the road for some groceries in your dad's pickup truck, for example. A delivery to the other end of town using only surface streets. A bus full of schoolchildren carrying our tiny humans to their respective homes or learning centers. It's hard to imagine that these kinds of brief, routine, minimal mileage trips in diesel-fueled vehicles are the worst for our ***only*** existing, widespread stopgap measure against black carbon pollution at the consumer level, but unfortunately, that is the reality.

Developed through exhaustive (no pun intended) research by our team at HNO Green Fuels for over a decade, the LeefH2 takes a much more direct approach by preventing at least half of that black carbon particulate matter from ever being generated during combustion cycles in the first place. By extracting hydrogen and oxygen from regular water

(H2O) and rerouting that hydrogen into the combustion chamber of a fuel engine, we are able to optimize the burn of that combustion flame and minimize the resulting residual unburned fuel (i.e., black carbon particulate matter). Our device is also maintenance-free and can even improve the maintenance cycle of diesel particulate filters (if the engine even has them … the larger engines do NOT). Our team has worked tirelessly to take any and all excuses off of the table to make this solution as accessible and widely adoptable as possible.

Backed by fourteen US patents, three Chinese patents and two Japanese patents, some of the proven reduction results for our LeefH2 technology include:

- 50%–80% reduction in black carbon particulate matter (PM)
- 12%–25% reduction in fuel consumption (and increase in cost savings)
- 24%–35% reduction in carbon monoxide (CO) creation
- 20%–30% reduction in nitrogen oxides (NOx)
- 45%–60% reduction unburned hydrocarbon (HC)
- 12%–20% diesel fuel usage reduction *while idle.*

(While the first five bullet points are crucial factors to benefit the fight against climate change, that last bullet point is especially important for vehicle performance, longevity, and fuel economy because traditionally, diesel engines are their most inefficient while idled.) The LeefH2 can provide diesel engine manufacturers and fleet owners or operators the ability to increase the life of vehicles and engines, improve their fuel economy, provide better protection for the environment, help businesses stay ahead of the curve of government regulations, save owners thousands of dollars in repair and maintenance costs, and provide one massive environmental bonus: creating, venting, and releasing net-new breathable oxygen into the atmosphere. Wait. Did I just say that this leafy-like device releases oxygen ***while*** improving fuel economy ***and*** eliminating black carbon? Talk about a triple threat.

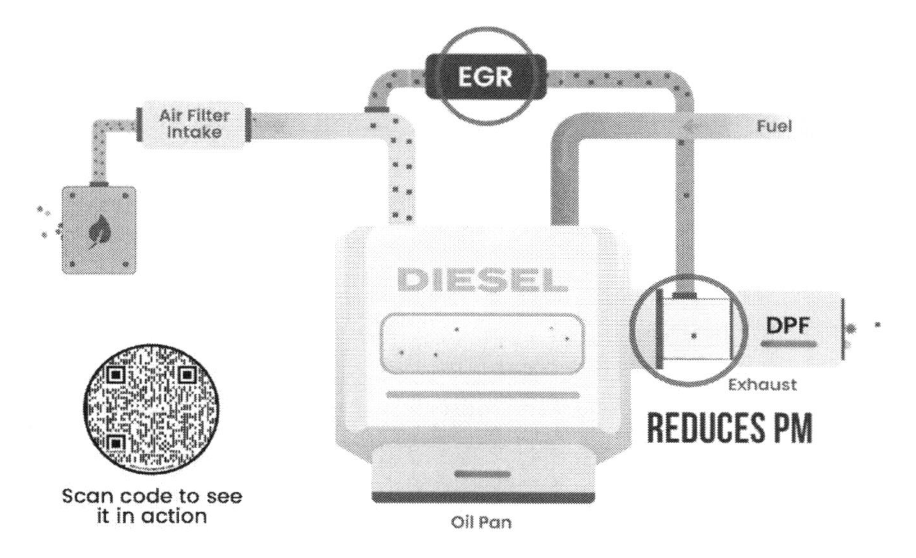

Alongside the true wizard behind the curtain, Webb Beeman, I spent the better part of ten years creating the core of the LeefH2 technology. The last four years were spent making the LeefH2 a mass-producible, commercially viable product. That process to develop and commercialize was neither smooth, linear, nor expected.

Originally, I'd just set out to create a device based on models and statements I'd read in a $69 book that I'd purchased off of the Internet in the hopes of dramatically improving fuel economies and reducing costs to the consumer. I was keen on saving a few bucks (and possibly making a few bucks while helping others save on fuel in the same way). I go into this whole discovery saga in greater length in the final chapter if you're curious to peek ahead and see how this remarkably unexpected solution came to be. **For now, though, it's simplest to just summarize a few key facts:**

- As a lifelong tinkerer, engineer, and inventor, I have been blessed to be able to find simpler solutions to complex problems than I'd ever imagined being capable of resolving. Something within me causes me to mentally and physically deconstruct something that I find interesting, explore the mechanisms and systems within it in an attempt to better understand how it functions, and then toy around with designs to optimize its operation, efficiencies, and ultimate applications. I guess I've done this for as long as I can remember, across engineering, law, software development, and other entrepreneurial industries and endeavors. And most of the time, the highest and best application of my inventions or developments has been found to be wholly different than I'd initially expected. That's just how life is sometimes, though—we can work so hard to find or force a solution to a problem that we can often totally miss the fact that we're already holding on to the antidote.
- The core LeefH2 technology has been developed over a decade of relentless deconstruction, optimization, tinkering, and testing. It is also not at all what it was originally intended to be. (*It is so much better.*) I've found that, in life and especially in business, the true testament of wisdom is being able to leave yourself open to the possibility of being totally wrong if it means that your work and your impact can ultimately be much more "right" in the end. The patenting of the LeefH2 helped teach me this lesson in wide (and truly unexpected) ways.
- The current version of the LeefH2 device was originally inspired by a tiny, plastic, clear cube-shaped electrolyzer—ever so slightly bigger than a Rubik's cube—that I had

discovered during the course of my research. It was being developed and sold by a German manufacturer that had built it strictly for educational purposes to help people visualize the process of pulling hydrogen gas from water. And, indeed, it was educational. Though my intended use for it was not at all how it had been marketed by the Germans, it successfully planted the seed of imagination in my mind and offered itself as a workable blueprint to build from.

- Soon after we started tinkering with that original German electrolyzer model, we discovered that it was far too fragile and would often shatter in the lab, so we knew it would never hold up in the hands of a consumer or in the rough-and-tumble world of mobile mechanized vehicles. If we had any hope of ever commercializing this thing, we needed to enhance its structure, strength, and durability. We knew we had to toss that original model out completely and start from scratch based on all that we'd learned up to that point—and I had no idea initially where to go next to properly iterate the device for our intended uses. Fortunately for me, however, Webb showed up in my life a couple of years earlier. He had already been mentally and physically deconstructing that device we were using to determine the optimal characteristics and conditions for it to work best, but his mental and physical deconstruction and optimization was like mine on steroids times ten thousand. Webb has one of those truly brilliant, visionary human minds I alluded to at the start of this book. I discovered over time that he could just listen to what I wanted to do and literally take anything that I had in my mind and build it with his hands. And usually in a mere matter of days. It was through Webb's genius and his ability to see all of the flaws of what we were using at the time that the first non-manufacturable version of the LeefH2 was conceived. After many, many iterations where each iteration was built by Webb by hand, my team was then able to take one of his working models and reduce it to four parts that could be constructed using an injection molding technique. A feat that also should not be underestimated. I also have to thank the genius of John Young (someone else who showed up years earlier) for making that happen. I wouldn't have been able to do that if I had been given one million years to accomplish it. Even 999,999 years and 364 days later, I would have been in the same spot. Thank God for Webb and John and boom, the first mass-production version of the LeefH2 was born.

- The current model is a small, rectangular, pancake-shaped device that holds a small quantity of water in a reservoir. When charged with a small amount of electricity, the system splits the water (H2O) within the reservoir into hydrogen (H2) and oxygen (O2). The hydrogen is then routed to and mixed with the air in the engine's air intake and speeds up the fuel combustion *in the combustion chamber*, thereby allowing the fuel to burn better (i.e., more completely) and create half as much harmful black carbon (i.e., toxic unburned fuel) as is produced through traditional diesel fuel combustion. While this is happening, the stream of oxygen that was split off from the hydrogen in the water is routed and released out into the atmosphere, thereby distributing environmentally positive, breathable oxygen into our air—effectively contributing a double whammy of positive outcomes in the context of climate change.
- The device is modular and can, therefore, be adapted for and integrated into any number or variation of fossil-fuel-burning devices to create virtually any quantity of hydrogen for any engine size or type. The applications are literally endless. While we are targeting diesel engines first and foremost (since they create black carbon on a critical-mass scale), the LeefH2 was also found to optimize the fuel consumption of

regular gasoline engines by up to 25 percent while conducting an EPA-standard Highway Fuel Economy Driving Schedule test, thereby creating a proven bridge and future opportunity to eventually distribute these devices across the consumer market, in addition to the heavy machines and diesel fleets we currently work with. So, while one size does not fit all, we have fine-tuned a scalable system to be able to build the proper size and number of LeefH2 units to optimize any and every existing engine on the face of the earth. The onus is on us (meaning humanity) to band together and push this technology forward in such a way that we can coordinate true, global adoption and, as a consequence, visible reduction in black carbon and a slowing of climate change. We have thus far only captured a small fraction of the entire universe of applications for the LeefH2 technology.

- We have over nineteen patents for this technology, with the most recent patent capturing the fact that this device can be universally adapted to augment literally any and every fossil-fuel-burning engine that exists (see above), with a first focused push that targets some of our world's greatest polluters—diesel engines. The previous patents capture the technology itself—the splitting apart of hydrogen and oxygen from water, before we were able to optimize it to become universally adaptable. That most recent patent is crucial because it allows us to scale this technology globally and actually make a true, tangible, real-time worldwide impact to eliminate the black carbon that is smothering us and our planet.
- The physical footprint of the LeefH2 device is astounding in comparison to the size of the engines or machines it can be adapted to fit. Currently, for large ships (approximately the size of a building) that are carrying something like twenty thousand shipping containers, the device size needed to service that engine is no bigger than a three-drawer filing

cabinet. Webb has already begun developing model iterations that could shrink the device down to the size of a cell phone for smaller engines. It is miraculous.

- The LeefH2 takes the guesswork and soul searching out of climate conservation because it will burn your fuel better automatically. You won't have to think or wonder or worry about how to better conserve or reduce consumption since this technology enables that automatically.

And those are just a few of the key facts, specs, and proven outcomes regarding our LeefH2 technology. When I first discovered that $69 book and its rough outline of a homemade fuel-saving hydrogen-based device—when I chose to pursue a path of tinkering around with that device to actualize the optimization and cost-saving promises that were made within it—I had literally no idea or dream that my life would quickly pivot and venture down the path of fighting harmful black carbon while creating a solution for climate change. I was not planning for it but, man, am I grateful that that's what the universe had in store for me. When I think back to the engineering roles, the patent law work, the Internet software development and entrepreneurial gigs I juggled all of those years, I could not be more grateful that this solution was what I was working so hard toward all that time. ***This*** solution. ***THE*** solution. It found me and presented itself to me unannounced but hugely welcome. It's almost as if the technology chose me, an endlessly curious, lifelong tinkerer who just so happened to have a background that made refining its form and function possible and to help it achieve its highest purpose, which can be leveraged today to help us to preserve our natural resources and save the planet (right here and now) as we continue to need and rely on fossil fuels as a global society to thrive within it.

We have identified the true enemy of the people and the planet. And now, the solution to climate change is ***here***. This is the part of the movie where we as human beings join together and amass all of our collective resources around our secret weapon and get to work to save our home from the alien species that's invading it. In this case, it is black carbon. This is the part where ***we*** become the solution, as well. So … LET'S GO.

CHAPTER SIX

Targeting Everything That Burns Fuel

"We have lived our lives by the assumption that what was good for us would be good for the world. We have been wrong. We must change our lives so that it will be possible to live by the contrary assumption, that what is good for the world will be good for us. And that requires that we make the effort to know the world and learn what is good for it."—Wendell Berry

In order to truly capture the depth and breadth of applications within which the LeefH2 can be leveraged as a solution for climate change (along with future iterations and improvements of the existing technology), we must first understand the scope and scale of the systems, products, and processes at play here. We need to first peek into those contributing factors that ***must*** be addressed in any reasonable, comprehensive, and effective environmental plan of action moving forward. Which, in the simplest and most abstract terms, ***must include targeting anything and everything that burns fuel.***

According to recent research released for the year 2020 by the US Energy Information Administration (EIA), approximately 60 percent of the energy produced within the United States was drawn from fossil fuels, with renewable and nuclear energy production sources gaining a great deal of traction over the past decade or two.[39]

On the flip side, however, as far as domestic energy *consumption* is concerned, the EIA estimates that about 80 percent of American energy use in 2020 was sourced directly from the burning of fossil fuels in various forms.[40] This basically means that though our modes of energy production have evolved to incorporate more and more "clean energy" options (we still have a ***long*** way to go, but hey, we gotta start somewhere)—our energy consumption has not yet matched pace with that change. As a civilization, we are still largely rooted in our fuel-burning behaviors, with many political and environmental action plans failing to address this stark disparity at the individual level. And we are not alone in this metric, as many other industrialized and emerging societies rely even more heavily on the burning of diesel engines and fossil fuels to charge their economy than we do. Natural gas, petroleum, and coal essentially power our planet, with current "cleaner" energy sources (which include nuclear electric power and renewable energy drawn from geothermal, solar, hydroelectric, wind, biomass waste, biofuels, and wood) trailing significantly further behind at about a sum total of 21 percent of overall consumption.

What is the biggest variable between fossil fuel production and consumption, you might ask? **The answer is: *the individual.*** If we see our large-scale production efforts (meaning, our manufacturing, energy farming, and enterprise-level power plants) starting to more actively lean away from the use of fossil fuels while our domestic consumption remains fairly stagnant on the higher end of the spectrum, we can infer that a great deal of that consumption relies on the choices, resources, and behaviors at the individual level. When we see our global population hovering around 7,800,000,000 human beings at this point in time, it becomes easier to recognize how crucial a part the individual plays in clean energy adoption and/or the eradication of black carbon in our atmosphere.[41] If we implement any macroeconomic change at the microeconomic level and then multiply that change's adoption by 7.8 billion, we bear witness to the behemoth of individualization in contrast with industrialization.

We said earlier that any climate plan of action that fails to address black carbon is merely an exercise in pissing in the wind. ***Now, it's worth stating that any game plan that fails to truly and measurably address***

and incorporate fossil fuel consumption at the individual level in its attack against global warming will also be doomed to fail at scale. This fix must first start with education, which means we must begin to better understand and explain to the layperson what exactly those products and processes on earth that burn these fuels actually are. The sources may be less obvious than most civilians would like to believe.

Take plastic, for example. Though this book is largely focused on black carbon's role and impact on our rapidly warming planet, the production and consumption of plastic in America and overseas plays an enormous and unignorable part in our weakening climate. And, while a great deal of effort has been put into place (especially domestically) to combat existing plastic waste, provide sustainable alternatives, and subsidize recycling options, we do not have as much visibility into the effect the mere production of plastic has on our environment. According to Christopher Joyce, an NPR scientific news correspondent:

> *"Plastic actually has a big carbon footprint, but so do many of the alternatives to plastic. And that's what makes replacing plastic a problem without a clear solution. Plastic is just a form of fossil fuel. Your plastic water bottle, your grocery bag, your foam tray full of cucumbers ... they're all made from oil or natural gas. It takes lots of energy to make that happen ... First, there are gas leaks that occur at the wellheads [where it comes out of the ground]. Then there are leaks from the pipelines that take oil and gas to a chemical plant. Then there's the lengthy chemical process of turning oil or gas into raw plastic resin. Factories then use more energy to fashion the plastic into packaging or car parts or textiles. Trucking it around to consumers generates more emissions. And once plastic is used, it often gets burned to make electricity, which is yet another source of greenhouse gases."*[42]

You see, we (human beings) are a population that is very much emotionally and physically attached to our plastics. Take a quick scan around your house, your city block, your grocery store or even your group paddleboarding outing at the beach, and you'll see mountains of evidence of this deep attachment to plastic everywhere you look. You'll see it out in plain sight—from the scattered fragments of man-made debris littered along our shorelines to the toys and tennis shoes on our playgrounds to our fridges at home that we keep fully stocked with plastic-packaged produce and products—while knowing full well that much of our plastic consumption has seeped into the hidden places (like within the stomachs of sea turtles, deep into the seafloors, and above ground in the tiny microplastic particles that circulate treacherously unseen within our breathable air).

Plastic has proven positively poisonous to our oceans, our land, and the multitudes of organisms and creatures that live within each, and yet, its use remains ubiquitous. To paint you a clearer picture in keeping with our theme, the smarties at *Stanford Magazine* have captured the statistics around plastic pollution rather bluntly:

> *"According to the EPA, approximately one ounce of carbon dioxide is emitted for each ounce of polyethylene (PET) produced. PET is the type of plastic most commonly used for beverage bottles. Other sources pin the production ratio of carbon emissions to plastic production closer to 5:1. Worldwide, we consume approximately 100 million tons of plastic each year. From the EPA's more conservative estimate to the more liberal one, that's anywhere from 100 million tons of carbon dioxide emitted to 500 million tons. With the more conservative estimate, plastics are on par with the annual emissions of 19 million vehicles, a number of drivers equal to the entire population of New York state. You can do your own car emission comparisons with the EPA's energy and emissions conversion calculator for annual passenger vehicles emissions. The liberal estimate puts us closer to the emissions equivalent of 92 million vehicles, or the number of drivers equal to the populations of every state west of the Rockies and Texas. Put another way, our passion*

> *for plastics resulted in emissions ranging from 10 to 45 percent of the annual emissions from the approximately 200 million licensed drivers in the United States."*[43]

Much like the larger argument contextualizing our reliance on fossil fuels, I must once again articulate a harsh but important truth: Every single modern human—from the staunchest environmental advocate to the most blissfully unaware or uninvolved civilian—uses or leverages plastic at least at some point every single day. ***Every single one of us, every single day.*** This is important because, in addition to damaging the oceans and littering our landfills, the demand for plastic's very use (again, by all of us) triggers an unavoidable chain of events that causes the burning of fossil fuels, and consequently, the creation and release of enormous amounts of destructive black carbon into the atmosphere, into our lungs, and onto our glaciers.

Whether sparked by the assortment of plastic bottles of water, juice, or our favorite soft drinks that invariably land in our hands daily—or by the plastic packaging that coats, covers, insulates, and separates almost every existing consumer good on the market—that consequential chain of events that ultimately leads to the production of black carbon looks something like this:

- The event in which one of *potentially hundreds* of employees who work at the factory that manufactures that plastic item drives to work each day in a (most likely) fossil-fuel-burning vehicle after leaving their fossil-fuel-burning home.
- The event in which said factory must be internally heated and cooled by fossil fuels to ensure that employee is comfortably able to operate heavy machinery.
- The event in which that heavy machinery, along with quite nearly every other output of energy inside that factory (lights, outlets, security systems, elevators, etc.), is powered and operated by electricity derived from fossil fuels.

- The event in which more plastic is used to package and protect that plastic item for safe shipping and delivery from the manufacturer.
- The event in which the trucks, trains, planes, or ships (big, small, and otherwise) that deliver those plastic goods to their retail or residential destinations run on large diesel engines and therefore contribute a grip of black carbon pollution in their wake to make the trip.
- The event in which one of potentially hundreds of employees who work at that ultimate retail destination (grocery store, supermarket, superstore, mall, etc.) that sells that plastic item must drive to work each day in a (most likely) fossil-fuel-burning vehicle after leaving their fossil-fuel-burning home to operate fossil-fuel-burning forklifts and other heavy machinery in order to stock, sort, and store that plastic item.
- The event in which potentially hundreds or thousands of customers who shop for that plastic item must drive to that retail destination in their fuel engine vehicles to purchase it (or, even more likely during the pandemic era, have it delivered by a fossil-fuel-burning mail or delivery service truck to their fossil-fuel-burning home up to several times per week/month).

Are you beginning to see a pattern here—or shall I continue?

And that's just a fraction of the use case specifically for plastic within a retail environment ecosystem. That doesn't even tap into the infinite and indirect ways in which we interact with and consume plastic (from hospitals, hotels, schools, and business meetings to sporting events, amusement parks, airports, etc., on and on ad nauseam), with plastic being just one of many, many black carbon contributors.

Say we want to look at cell phones instead (most of which also happen to be made up of a great deal of plastic, but that's not my point here). Our cellular devices and Internet technology have become so ingrained into our general mode of being and communicating that we have seen newer commuter vehicles come equipped with a "Check Your Backseat" alert that gets triggered after the car has been turned off if weight can be detected in the backseat—because more than one news story has shown us that parents are sometimes more likely to remember to take their cell phone with them out of their hot car than their children or pets. Many of us have "Screen Time" stats that reflect daily device use at every waking hour (and sometimes interrupting our sleeping hours, as well).

Do you ever wonder how that cell phone got to you in the first place? Have you ever paused to consider all of the processes, people, and products that remain in motion to keep that cell phone operating without excessive flare-ups, connection failures, or outages? In most cases, the answer to each of those questions very closely mirrors the chain of events we just outlined for plastic. At the highest level, it at least requires data centers (which would operate on fossil fuels and employ those who likely drive fossil fuel engine vehicles), computer servers (which also leverage fossil-fuel derived electricity), cell phone towers (which not only operate on fossil-fuel derived electricity, but require large diesel engine trucks, rigs, trains, or other massive black carbon contributing machines in order to be shipped to, dug out, and constructed at the proper location after departing their fossil-fuel factories), and maintenance teams (who may or may not stop off at a 7-Eleven convenience store to pick up a quick bottle of water or juice—also plastic—while en route to the data center or cell phone tower in their fuel engine company vehicles from their fossil-fuel dependent offices or homes).

If you can't tell already, this thought experiment could go on forever. I'm exhausted by my own explanation already, and that's just a ***tiny glimpse*** into how deeply reliant modern civilization is on the products and processes that require the burning of fossil fuels to operate. If we think we can get anywhere close to being "100 percent green" any time soon without leveraging a solution that can optimize this fossil fuel usage, we are sorely mistaken, foolishly kidding ourselves, and deeply missing the extent of the issue at hand. ***Which, again, requires a sharp***

focus on drastically decreasing (and ultimately, eliminating) the creation of black carbon particulate matter from anything and everything that burns fuel already.

And that population sample of "anything and everything" that burns fuel is rather vast. In the United States alone, we have about 7,658 active power plants that rely at least to some extent on the consumption of fossil-fuel energy to produce new energy. According to *Arcadia*, "Coal plants are a leading source of carbon dioxide emissions [and black carbon creation], accounting for 1.7 billion tons [in a single year]."[44] And still, in the estimation of Caroll Muffett (head of the Center for International Environmental Law), the harmful emissions from the production and incineration of plastics alone between now and the year 2050 will surpass the annual emissions from ***all*** of the coal power plants in America ***almost fifty times over***.[45] (A friendly reminder of why blaming and bullying the fossil fuel industry isn't the way to solve this problem, nor is it even the proper context within which to discuss it.)

Did you know that one of the primary global contributors to harmful CO2 emissions and black carbon creation is stovetops? Like, the ones used for cooking food for families where a microwave or fancy, stainless steel range oven simply aren't options? Meaning, those makeshift devices leveraged for survival by countless humans in some of the most under-resourced areas of the world? Do we truly think we're going to be able to provide and/or subsidize a "cleaner energy" replacement to this fundamental necessity for millions of people at scale, and anytime soon? Who knows though, maybe we as humans can come up with a LeefH2 version that can use water to create the hydrogen fuel that can be used for cooking? Human beings … anything is possible.

What about the millions of fuel-operated backup generators in rural counties and developing countries that contribute a huge chunk of black carbon and harmful emissions while literally sustaining human life and business and economies in the hardest to reach parts of the globe? According to a report released by the International Finance Corporation (IFC) of the World Bank Group, "in Western Africa, for example, the electricity provided by back-up generators is equal to 40 percent of the electricity generated by the grid."[46] So, do we allow the fact that these "generators are contributing significantly to the emissions of fine particulate matter (PM2.5), sulfur dioxide (SO2), nitrous oxides (NOx), carbon dioxide (CO2) and other pollutants that compromise human health and contribute to climate change" negate the fact that they also directly power the most basic conveniences and functions necessary for survival? My answer: no. Because we don't even have to entertain that kind of Sophie's choice. ***We have a solution, here and now, that can ultimately help them burn that fuel better.***

The point isn't to dwell in the challenges we are facing; it is to delight in the solution: the LeefH2 technology and every possible iteration of it that can augment, enhance, and ultimately help burn the fuel better of every existing and future fossil fuel engine, product, or process. Are you starting to see it? Good. Because we need you. You and the rest of all of the incredibly creative, solution-oriented, problem-solving, no-task-too-large human beings that can start seeing and developing a solution for "everything that burns fuel." ... everything. From the smallest mopeds in India to the largest container ships in the Suez Canal to every type of aircraft, from helicopters to jets.

Black carbon is the enemy. We need to find a way to collectively reduce and eventually eliminate it, or we are most certainly barreling toward a destiny that will no longer include a beautiful planet that is truly unique in the universe. A wonderful blue planet with snow-capped mountains, vast blue oceans and glaciers on both poles, lush green forest and jungles, and countless valleys where we can grow food and live. That's the planet that I hope we all want to keep.

CHAPTER SEVEN

Burn Fuel Better

"Once you have an innovation culture, even those who are not scientists or engineers—poets, actors, journalists—they, as communities, embrace the meaning of what it is to be scientifically literate. They embrace the concept of an innovation culture. They vote in ways that promote it. They don't fight science, and they don't fight technology."—Neil deGrasse Tyson

At this point, I'd say we can probably all agree on one thing: Our climate is in crisis, and today, in 2021, we can hear the siren calls screeching through every single weather forecast, through every historic natural disaster, through every news story and new medical research release. We can see the evidence of this crisis in every single uptick in temperature, just as we can see it darkening the glittering surfaces of our polar ice shelves. The destruction is visible everywhere—and it is now abundantly clear that something must be done here, at home, and immediately.

With the new administration we have in office there *have* been a few really fine environmental action plans put into place already. According to a December 2020 study released by Princeton University, several avenues toward carbon neutral (and best-case-scenario, carbon-negative) outcomes in the United States have been brainstormed, blueprinted, and bylined. These scenarios are both sophisticated and idealistic, but every single one of them requires us to put multiple major moves in motion well before 2030, including "putting 50 million electric cars on the road and 3 million public charging ports, increasing the use of electric heating systems in homes from today's 10% to 23%, tripling the use of electric heating on commercial property, quadrupling wind and solar capacity from today's 150 gigawatts to 600 gigawatts, building high voltage transmission infrastructure to carry renewable energy over long distances and reducing non-carbon dioxide (CO_2) greenhouse gas output, nitrous oxide (N20) and methane (CH_4), by 10%."[48]

However, while those aggressively planned pushes are indeed possible through a major collective effort, ***they will simply never solve our global warming crisis in time or sustainably without also directly addressing the issue of the treacherous villain of our discussions thus far, which is the toxic black carbon particulate matter that's poisoning our planet.*** Since all of these existing global action plans still require the burning of fossil fuels to truly take flight, we will forever continue to churn out black carbon as a by-product of even the greenest calls to action, and those black carbon particles will then continue to drift down onto our glaciers, heat our atmosphere, and weaken our timeline for human life on earth. This beast of black carbon must be conquered now if we have any hope of continuing life on this planet as we have grown accustomed to over the past several centuries—yet none of these existing plans of action address black carbon in any meaningful or measurable way. Until now. Today. And that solution is surprisingly simple. We must simply start burning those fossil fuels better by leveraging hydrogen-powered technologies like that of the LeefH2.

Truthfully, I would almost guarantee that every world leader, politician, industry influencer, and business tycoon is not likely to be aware of the facts outlined within this book. Especially concerning black carbon. Even if they are aware of black carbon like most scientists are, up until now, there was nothing NO ONE could do about it. What if we recognized the actual enemy for what it is (the creation of particulate matter and black carbon that every single one of us contributes to but that we can all actively work together to resolve), thereby creating one global, concentrated, collaborative effort to smash climate change beneath our feet and save humanity's home for a few more millennia? What if we set our differences aside and joined forces instead of muddying up the narratives and pointing fingers? What if we identified black carbon as the alien species that looms just outside our atmosphere and threatens to invade Earth through cryptically coded messages fielded and figured out by the smarties at NASA? I mean, how do we typically see that scenario play out in films? Do the countries start bickering and turning against one another and allow their own internal differences to tie up their attention while the aliens just have their way with the humans? No. Of course not. Nobody would watch that drama. Instead, we line up and fill up the rows of movie theatres with our popcorn, candy, and soft drinks (ahem, more plastic, ahem) in hand, eagerly anticipating the moment in which the whole world bands together to cohesively fight the good fight—because they have one fundamental, immovable thing in common: They are all humans at war with something inhuman. And so are we, here. Humans may be at fault for climate change, but we are also uniquely capable of crushing it before it crushes us. So, you must just ask yourself: Do you want to save the planet, or not?

We are literally standing smack dab at the precipice between a global climate disaster and the opportunity to give humanity a chance—some extra time—to resolve some of the problems we have created during our stay here. While humans have evolved to exceed even the wildest stretches of our imagination, we have simultaneously created an inorganic, biological toxic waste product (black carbon) that our environment hasn't been able to match pace with and evolve alongside to break down or absorb or digest as quickly as it is created. And it is our consumption, our tireless demand for our comforts and conveniences, that is creating this malignant, toxic waste as a by-product.

We didn't know that things would happen this way. We couldn't have known, and I'd like to believe that if we had, we might have done things a little bit differently. But we didn't realize the consequences of our actions—and even when we first grew suspicious of the environmental threats at play here, we started (and have continued) targeting all of our attention and resources toward the wrong enemies. We need to now evolve our approach, along with our thinking. As Einstein once said, "We cannot solve our problems with the same thinking we used when we created them." That next level of solution-minded thinking requires us to step beyond the misconceptions of climate change, beyond the blame game and the pointing of fingers from environmentalists toward "Big Oil" and back, beyond the belief that we could ever truly stop burning fossil fuels so long as we wish to remain even partially civilized. We must instead categorize black carbon as the alien species set to invade and destroy our planet, our people, our home—and label that invasive alien species as the true enemy and greatest threat to our survival. We must also do so with the startling self-awareness that we (every single one of us) have created that enemy in all of our endless tinkering and brilliance and inventive feats of the imagination. We must acknowledge and accept that this is an enemy whose warpath and destruction we are fully responsible for remedying. In the face of that enemy, all politics must disappear. All language barriers, ideological differences, and divides between classes and caste systems and cultures must dissolve. If we can do that, we can do this. We can band together and leverage this remarkably simple solution today. We can get organized, collaborate, and cooperate together to fight back against that alien enemy's attack and take back control of our planet. We can breathe new life and infuse fresh oxygen into it. We can heal it. We can make of it an entirely new home. In turn, we can become our own solution, over time, rather than the key contributor to our own demise.

While this may sound like a tired plotline for a summer blockbuster movie, it is indeed our reality. Today. It is happening right now. The fight is already well underway. People are dying. Lungs are collapsing. Glaciers are melting and breaking off from the ice shelves. Vegetation and wildlife are decaying under the thick, dark dust of black carbon. And time is running out. The time is now. The time is right. This is a fight for every single one of us, and it is a fight that can only be won if every single one of us joins together in our attack against the enemy. We decide what happens next. ***WE DECIDE***. We can make the choice and take collective, collaborative action to cut out the creation of black carbon, right at the source during the initial fuel combustion process, today. And we can do so simply by burning those fuels better.

CHAPTER EIGHT

New Technology = New Global Economies

"Every once in a while, a new technology, an old problem and a big idea turn into innovation."—Dean Kamen

So what do you think would happen if the entire world got together to fight the common enemy of black carbon? You think it would have a positive effect on the economies of the world? This is such a massive undertaking that it is not only possible for the world to benefit environmentally but also economically. Every county in the world has and will continue to have issues with the burning of fossil fuels. For ANYONE to think that it will simply go away, they are truly living with their heads in the sand. It will not go away with the proliferation of electric vehicles or hydrogen fuel cell cars or solar panels or windmills. It will persist during and long after any attempt at a conversion to green technology. It will ALWAYS take fossil fuels to make any and everything "green" for our conversion to green technologies. Maybe I should repeat that another way: ***It will always take the burning of fossil fuels even for the hoped-for conversion to a totally green energy society. Always!***

As mentioned before, this is not an either/or situation. Reducing black carbon and green energy conversion has to happen simultaneously and over the next fifty years. The existing technology that has put our entire society in this pickle must be retrofitted to "Burn Fuel Better" and reduce black carbon emissions so that they can assist in the green energy conversion. If humankind can somehow manage to put greed aside and stop the finger-pointing, this can have a positive environmental and economic effect on every last human being on earth. Every last country and government must be involved. The applications are endless. When we say everything that burns fuel, we really mean *everything that burns fuel.*

If a totally green conversion must happen worldwide, so must be the worldwide effort to reduce or eliminate black carbon. Perhaps now would be a good time to understand the positive economic advantages and incentives that would ensue if a collective global effort to eradicate black carbon and particulate matter from the globe was to occur.

First, most of us within the realms of business, politics, engineering, and science already know and understand the grounding economic truth that whenever a massive technological or scientific disruption occurs at scale, a nearly infinite (and positive) economic ripple is released from that innovative epicenter of theory and design. Spreading out with speed in traceable, thriving pathways from that initial inventive spark, we can see the catalytic gains and unalterable impact from such innovation at both the macroeconomic (societal) and microeconomic (individual) levels. Simply put, new inventions breed new industries, and new industries breed net-new jobs, revenue streams, research, discovery, competition and, as a consequence of each, measurable economic booms and boosted political positioning. Innovation is power. Period.

Historically, academically, and conceptually, this proven economic pattern has traditionally been tied to scientific exploration and mechanical engineering. It has been applied to those things that can be built, designed, mass-produced, distributed, and exchanged. To those tangible artifacts and devices that can be held and grasped tightly on to. To those items that can outlive and outlast even those masterful brains and bodies which first conceived them. However, we have clearly seen that the twentieth century was an industrial revolution wholly and uniquely unto itself—with its final quarter alone (1975–1999) comprised of some of the most vast, incomprehensible, and intelligent growth and innovation humankind has ever witnessed in the entirety of its documented history. With the advent of the Internet, satellite engineering, mobile technology, and artificially intelligent design, we became a society that thrived and was driven forward by those invisible things that cannot be seen or touched, but that could evolve and propel us forward as a species in such a way and at such a rate that natural, biological evolution has never come close to matching.

When we then skim slightly to the right on the timeline of humankind and creation to gaze solely at the first fifth of the twenty-first century (2000–2020), with all of the machine learning, bioengineering, cryptocurrencies, digital spaces, 3D printing, and in-home robotics it has churned out with haste, we glimpse a potentially startling truth: We have already begun to exceed the farthest reaches of our wildest science-fiction fodder. Even in view of the COVID-19 pandemic—in which an organic, novel super virus threatened the basest element of our global humanity (our bodies, our fleshy operating systems, the remarkably vulnerable meat that we are all made up of)—we were able to almost immediately halt all in-person interactions and largely stem viral transmission globally by pivoting all production, interpersonal assembly, and general modes of being to virtual spaces and solutions. We were able to snuff out parts of the virus's warpath and protect the most vulnerable communities, while preventing a total collapse of our global economy. Sure, there have been significant losses, both in our populations and in our industries—but those losses are nothing compared to that which would have occurred had we not allowed advanced technologies to intervene and spare us unnecessary destruction. For the first time in human history, we were able to effectively outsmart a diabolical, biological enemy simply by leveraging our newest tools, tactics, and technologies that it could neither penetrate nor destroy. The inorganic went to battle against the natural state of the things, and it won. If 2020 is a testament to any singular truth, it is this: humankind (and our modern inventions) has successfully reimagined, redefined, and rewritten the rules of what is actually possible here on earth.

However, this will be a perfect time to remind everyone that all of these incredible steps forward by humankind on earth were made by burning one heck of a lot of fossil fuel. We couldn't have done ***any of it*** were not for fossil fuel and the burning thereof for all the things that are required for human existence. Warm or cool buildings, agriculture, mining, construction, shipping, Internet communications, cell phones, and so on. It was NOT based at ALL on electric vehicles, solar power or wind energy. ***Not at all!*** (does everyone understand and know what **not at all means?**)

But getting back to global economics, the theory of innovation economics, grounded in entrepreneurialism, institutional evolution, technological advancement, and enhanced productivity, is exactly the force that supported humankind through the deepest and darkest trenches of commercial and industrial lockdown during (and will continue to after) the pandemic. It is what encompasses future industrial projections and energizes active experimentation and design. It is the uncapped economic reality that this fuel optimization solution can create, here, today, within our lifetimes. You see, we do not necessarily need to create net-new replacements for our existing fossil fuel resources and systems in order to save this planet and thrive economically, as traditional theories might espouse—we simply need to innovate, augment, and enhance the current operations of those resources and systems to burn fuel better, as innovation economics purports.

And the proof of its efficacy is in the pudding, a million times over.

Take the iPhone, for example. While Apple's disruptive device managed to seamlessly combine and streamline the top features of mobile telecommunications, personal computers, cameras, music players, and web browsing technologies at its 2007 debut, it was still always a cell phone at its core. But the handheld cellular phone had already been invented decades prior by Martin Cooper, an engineer for Motorola, and iterated countless times since its initial inception.[49] Born in the midst of a competitive rivalry with his professional nemesis (Joel Engel of Bell Laboratories at AT&T), Cooper's hefty portable prototype made history in 1973 as he used it to make the world's first mobile phone call one April afternoon from the streets of Manhattan. This revolutionary invention was not without its challenges, though, as it was riddled with functionality and cost issues that seemed to discount it from scalable commercialization (does that sound familiar?). Motorola dedicated endless resources and ten years' worth of research and development into improving Cooper's original model and transforming it into a trimmer, consumer-friendlier version that sold for between $3,500 and $4,000 a pop. Now, remember, that cost in 1983 looks a bit closer to $9,500–$11,000 when converted to the purchasing power of today's American dollar.[50] That's a sizable chunk of money that would outprice the average consumer—and the size, function and sustainability of the updated design were still not ideal even at that high cost. It would likely have been difficult for Cooper to imagine that, not too far off into the future, quite nearly every civilized adult would carry one of his inventions in their purse or pocket. Of course, thirty-four years had passed between that first prototyped mobile phone call and the launch of the first iPhone, but I would argue that Steve Jobs's creation was less a novel invention and much more a feat of reimaginative innovation. It enhanced how mobile phones were leveraged, packaged, and even perceived by introducing technologies and new means that could accomplish its initial ends better.

By disrupting one industry, the telecommunications field, the iPhone also consequently spawned or supercharged several net-new consumer-facing industries, including mobile video conferencing (FaceTime), integrated personal messaging and communications (iMessage), cloud-based data storage (iCloud), social media (lest we forget that Instagram was originally an iOS-only platform), music collection (iTunes), pocket artificial intelligence (Siri), GPS positioning (Maps), vertical news storytelling, and, perhaps most substantially—the mobile app, gaming, and swiping economies (which, in turn, have sparked unicorn transactional brands such as Uber, Lyft, Airbnb, DoorDash, Tinder, Netflix, Spotify, Twitch, Venmo, and the like). And by that measure, the iPhone would be a stunning example of innovation economics at play. Considering that the iPhone alone, separate from these subindustries it helped create or catalyze, managed to sell 1.2 billion iPhones (worth $738 billion) in its first ten years, I'd say there is something to be said for the quantifiable, visible value of innovation economics over traditional invention.[51]

To that same effect, the mass-production and implementation of fuel optimization devices and systems, such as the LeefH2 technology, to burn fuel better can positively disrupt the fossil fuel *and* green energy industries while spawning new industries through its ultimate iterations, without forfeiting all of the existing resources currently at play throughout our precious planet.

Consider this: Theoretically, if we scaled out the fundamental technology that forms the current LeefH2 device to the full spectrum of applications, it could ultimately serve in helping us burn fuel better as a solution to climate change. Imagine how many jobs, companies, industries, and economies would be created as a result of that collective effort. Endless engineers, manufacturers, installation and maintenance technicians, designers, developers, salespeople, marketers, advertisers, analysts, regulators, resellers, educators, legislators—the list could go on. And, considering that global warming is not only an American issue, the economies and industries that would be sparked and/or impacted by iterations of this innovative technology would reach every single corner of the world. It would reach those multiple millions of backup generators, diesel-engine delivery trucks, military vessel fleets, third-world stovetops, and air-conditioning units. The possibilities for global economic gain here are limitless, but they are merely the cherry on top of the central, sweet ice-cream sundae that is: preserving our planet against its most poisonous polluter (black carbon particulate matter) long enough for us and our offspring to continue life on earth with any degree of civilized normalcy. But hey, why not have both. Both's better. After all, I do always ask for the cherry on top.

And those are just a ***fraction*** of the potential global economic implications that stem from this game-changing multiple-choice climate change dilemma. Based on this information, which plan of action and consequential outcome would you choose? Could you even make a choice? (Might I suggest the final bubble, which is in fact, all of the above—invest in renewable green energy reasonably and responsibly while also burning existing fossil fuels better through hydrogen electrolysis solutions, such as the LeefH2 technology. No jobs lost. Scalable solutions. A future guarded for humankind, our planet, and all of the various flora and fauna that comprise it. Everybody wins.)

Now, I realize that by inciting economic consideration of this technology—my technology—I am simultaneously inviting competition. And I am doing so on purpose. Because, while the LeefH2 is today's dark knight solution for eliminating black carbon creation and therefore arming the global fight against climate change, that fight is much bigger than one army. That battle is certainly bigger than one commander is capable of leading. We need everyone to get involved and drive this innovation and its consequential economies to the most scalable heights possible. While I'll be working tirelessly throughout this life to bring this solution to the masses and curb climate change as best (and as quickly) as humanly possible, I am only one person. I have a dynamite team, but we are only a small representation of minds within this space. Therefore, I welcome any and every imaginative tinkerer capable of expanding, evolving, and enhancing my solution for all of our sakes—to capture every possible application, iteration, and revolution of this solution. I need everyone. I need ***you***.

Will you take up this fight alongside us?

Will you answer the call to save our planet, our home—our futures?

CHAPTER NINE

Who the Heck is Donald Owens?

"Bringing simplicity into our lives requires that we discover the ways in which our consumption either supports or entangles our existence."—Duane Elgin

I did not set out in life to help remedy climate change or eliminate black carbon from our planet. I hadn't planned to become an inventor. Or an entrepreneur. Or an engineer. Or a patent lawyer. Not at all. Not even a little bit. However, I (like the ***vast*** majority of you) have always hoped and dreamed of leaving behind a positive impact on this world. Maybe I could even live the kind of life that would give you a real reason to give a crap about what I've done or what I think. While it's hard to imagine how to accomplish this when you still see yourself as an ordinary guy just trying to do his best (like I and most of us do), life has a way of surprising you, and fate always has a way of finding you right where you are—ready or not. The challenge then becomes trusting the process and believing that you are exactly where you are supposed to be, doing what you are meant to do, for the highest possible purposes. But man, trusting the process ain't always easy, is it?

You see, when I was a child of about seven or eight years old, my dreams were not defined at all and felt more like everyday adventures. I didn't spend my days imagining I would become a fireman or astronaut. I wasn't a musical prodigy or math genius. I was just doing what little boys do at eight years old. And, perhaps because of that, I remember having a wonderful, uncomplicated, comfortable childhood. I had one brother and three sisters at the time, and I can remember that we traveled all over the place. As the second son to a family that would eventually produce one extra brother, for much of my life, I was the classic middle child. All of the attention went to my older brother and the eldest of my sisters, as did all of the perceived pressure. I didn't carry any of the weight or responsibilities that came with being the oldest son or the oldest daughter. And sometimes, it seemed as though I could get away with just about anything. And most of the time (not all the time), I did (mostly insignificant things, of course). I can remember doing all kinds of crazy stuff in high school, skipping classes to go and smoke weed, going down to the local "head shops" (where you could buy everything from jeans to marijuana pipes) and actually buying two pairs of jeans while stuffing another pair in my pants and walking out with three pairs. I have no idea what possessed me to do that, but it was fun and exciting to both myself and a couple of good friends of mine at the time. To me, it was somehow a thrill. However, I can vividly remember when I stopped doing it. One day one of my friends told me that another one of my friends got arrested and was put in jail. Out of sheer ignorance and naivety, my response was, "What? You mean you can go to jail for that?" Man … was I a dummy. I don't know why, but I had no freaking idea that actually "going to jail" could be a consequence of my actions. That and a few other things that I can think back on still make me realize how easily life could be so radically different. And how things could be so much more difficult just from one little and somewhat innocent mistake. That young people make all of the time.

Turns out, I didn't really know what I wanted to do at twenty, either. But I always had a knack for tinkering with things and a capless curiosity. Specifically, I had a knack for taking things apart more than putting them back together—because, once deconstructed, I could discover how they were developed, analyze what elements composed them, study how they operate, and learn from their fundamental framework. I built my first car, a '55 Chevy, when I was eighteen. Found a shell of a car, painted it, put in an engine and transmission, and fixed just about everything that went wrong with it as I drove it all over Florida. In hindsight, I see it as the early embodiment of my essence. My calling. I am a tinkerer.

So, at twenty years old, still without a clear vision of what I wanted to do with myself professionally, I took my tinkering as a sign that I should pursue a career in engineering. I started off at Florida A&M in a two-year engineering program where, part of the time, I worked at the Kennedy Space Center as an engineering student. It was a great experience seeing all the workings of the space program. At the time, I think, it was right after we landed on the moon.

Next, since A&M was only a two-year program, I discovered (actually my father discovered it and told me about it) an immersive university program at General Motors Institute (now Kettering University) in Flint, Michigan, where I could study for six weeks at a time, and then work for six weeks at a time, all while earning an engineering degree. By the time a semester was over, you would have some level of real-world work experience of the practical applications of your studies. I majored in mechanical/electrical engineering and wrote my thesis on plant steam distribution and usage. By the time I graduated, and having worked as an engineer for General Motors for several years, I knew, without a shadow of a doubt, that I did not want to be an engineer. It's not that I minded the role or the work. I just didn't like the limited/limiting environment that came with it. And, like many people, I was more motivated by what I didn't want in life than by what I did want. I was looking for a way out.

Fortunately for me, I had a friend in Flint whose cousin was also an engineer and had got into a program to work for Western Electric and to go to law school to study patent law at night. Apparently, a background in engineering is just the ticket to a career as a patent attorney. Not that I really wanted to be a lawyer, but I definitely did not want to continue as an engineer. This felt like the perfect combination of my technical tinkering with my desire to escape the structure of the engineering world. So, off to law school at Georgetown University, I went.

So in the late 1970s, I began writing patent applications and working for Western Electric and Bell Labs. By the time I graduated from law school (early 1980s), I had realized that, even though there was good money to be made in patent law, the focus felt too narrow for me. In order to write patent applications effectively, it's almost as though you need to think like a laser beam, and I somehow believed that I was born to think like a searchlight. By the end of the decade, I'd packed up and left the patent practice to pursue more creative, inventive, entrepreneurial endeavors. To tinker more freely.

Flash forward to a decade or two chock-full of small business ventures, multilevel marketing gigs, real estate projects, and various other ad hoc, independent occupational pursuits. I made some money in some of those efforts, not so much in others. As the Internet was first emerging, I had learned how to toy around with databases and computers, and I eventually began writing and developing DOS applications, accounting software for Windows, and Internet-based programs. I entered the technology field mostly out of a need to survive.

But, above all, I look back on that season of self-employment and hustling and remember it fondly because I was doing it all for myself and my family. I didn't have to answer to a commandeering boss or punch a time card or stroke anybody's ego at a corner office watercooler. I felt free—from the claustrophobia of engineering, from the tunnel vision I felt within patent law, from the confines and constraints of working for someone else to build their business, rather than my own. I could do anything and everything that intrigued me professionally, and I did. I'd spent all of my early adult years receiving the best formal training possible across industries while pushing back against expectations, traditional timelines, and predictable outcomes. Throughout it all, an inventing mindset remained and permeated through every single project I put on my plate. I was driven to develop new ideas and innovations, sure, but mostly, I wanted to find (or create) work that could maybe, someday, help another person's life (like most people reading this book).

While I was working in technology and computers, I realized that most of the accounting software solutions on the market at the time were just too complicated for most people. For most business owners. For most non-accountants. Even for me. It was all just too much. I remember watching my father as he struggled to keep track of his business tools and finances, and saying to him, "Dad, I think I can write something for you." So, I decided to create a simpler solution and began building out some accounting software that could actually work the way that most people worked. I would take the traditional, decentralized, paper-based processes I'd observed my clients, colleagues, and family used, then transition and translate those processes into computer programs that could organize, track, and manage financial information in an easier, more intuitive way. Meanwhile, the Internet was bursting our world wide open and forever altered the fundamental fabric of how we, as humans, could connect, communicate, and operate. Suddenly, on-premise applications were moving online and becoming available from anywhere, and I saw that as an opportunity to expand my software development skills and meet the business needs of my clients from wherever they were at. The digital possibilities were endless, and it was this work that kept me going and challenged me creatively as an entrepreneur through the new millennium—all the way up until one fateful evening I had at home in 2008.

By then, between business hours, I'd still been toying around with countless other ad hoc options for projects, items, and improvements I could make on my own in order to, ideally, commercialize and sell them to others.

One night, I stumbled across a short paperback book that was being advertised as a kind of instruction manual for how to build a small device that could fit into your car to help you save gas (and ultimately, save money). The book said that the device could take regular water (H2O), and run your engine on the hydrogen in the water. It was just a small paperback book, maybe only about fifteen to twenty pages or so, filled with illustrations for how to build what was essentially a hydrogen generator. It was selling for $69, but it'd piqued my curiosity enough for me to purchase a copy. I figured the book would pay for itself if the device could actually help me start trimming down my own gas usage, but more importantly, perhaps it could help me understand how to build something similar that I could sell to help others do the same. It seemed like something worth looking into.

When the book arrived, though, I must say—I was a little disappointed. The outlined instructions were quite inadequate, and the approach that the author had taken led to a fairly unworkable design in my opinion. ALSO, I could not find any of the materials or parts they pointed to in the book at my local hardware store. The design itself was OK (I guess), but I just couldn't shake the feeling that there was another possible approach to build the hydrogen generator even better. By that point, I was already hooked on the concept, so I began to seek out other materials to build it.

In doing so, I quickly discovered that I may have stumbled on a tinkerer's trend, as quite a few other people had also started building these hydrogen generators. While a commercial solution still hadn't been released yet, I started spotting more and more of these hydrogen generators that were being built in various garages and small shops all over the world. Many of the device developers were also making some rather outrageous claims about the technology, effectively saying that by implementing the device, you could increase your fuel economy by as much as 150 percent.

In other words, if your car was getting an average of thirty to forty-five miles per gallon before you augmented it with this kind of device, you could squeeze an extra forty-five to seventy miles out of every gallon.

Basically, they were suggesting that you could more than ***double*** your existing miles per gallon and surpass even the most impressive fuel economies on the market at the time—just by using regular water. The promise definitely sounded like it could add up in meaningful cost savings over time.

The thing is, though, I didn't realize that the projections were aggressive or outrageous (and maybe even impossible) at the time. All I knew for sure at that point was that I was intrigued enough by the idea of pulling hydrogen out of water and using it to boost the fuel efficiency of a vehicle that I couldn't help but keep tinkering around with the engineering behind it to see if it could work.

I was determined to build a workable prototype to see if I could make some of those claims come true. The book had given me the basis and framework to explore better variations of its central thesis—that a handmade device that extracted hydrogen from water could massively enhance fuel usage. I studied the thing for weeks. By then, I couldn't let it go. I was hooked, because the idea to optimize, if not double, a car's fuel economy had been planted in my mind. Someone had supposedly found some way to use water to power a vehicle, and I knew that I could find a better way to make it happen if I just kept trying.

So, I kept trying, to say the least.

After closely studying the book's instructions and gathering enough additional information from other sources to flesh out a fresh concept and structure for the device, I began building something in my garage that went beyond the scope of the book's outlined solution. I built and installed a makeshift device in my car and began driving around with it for days on end to test out its capabilities. I had developed some very crude ways of gauging fuel performance and mileage savings, if there were any to be found. I would drive all over the place, hundreds of miles at a time at all hours of the day and night, often as far as Las Vegas from Temecula and back again, in order to determine whether or not the thing was actually working.

Now, it's important to note here that, in 2008, we were still in the very early phases of hybrid or electric vehicles and smart automotive technologies on the consumer market. So, I was able to figure out a way to measure the efficacy and efficiency of my gas mileage that today's car computers definitely wouldn't allow me to do. In more ways than one, this was a blessing—a serendipitously aligned culmination of timing, invention, and opportunity.

My appetite for demonstrable data quickly became pretty serious. I would jot down any and every detail from each drive, with and without the device turned on, to log every possible change or fluke or discrepancy, so I could better understand what I was working with and how I could stabilize the system.

The goal of any theory or invention is to be able to replicate and repeat your findings with as minimal variations to the outcome or output as possible. So, I would try to simulate each test drive to reflect exactly the same patterns, mostly the same routes and road conditions, at about the same times of day, lengths of time, and miles per hour. I would always try to make the drive in the early mornings or late at night when there would be little to no traffic in the way that could muddy the variables or cause fuel fluctuations. The testing had to be as precise and controlled as possible. And every single time, the device worked.

I saw a minor but measurable difference in the gas mileage I was getting with the device than what I was getting without it. It wasn't a major improvement, with gains usually landing between 10 percent and 20 percent, but it was there, every time. I did, however, wonder, "Why am I not seeing 50–150 percent improvement (per the book's estimates)?"

After digging deeper into more research looking for an answer, I discovered what others had already learned about the so-called "hydrogen" that was being produced by this process: the resulting output was actually something called "oxyhydrogen," which is a combination of hydrogen and oxygen gas (or HHO). My car, like all modern vehicles, had "oxygen sensors" that measured the amount of oxygen in the air to determine whether the car was either running "rich" or "lean."

Since my prototype added extra oxygen to the system, the car's computer would automatically make an adjustment to the fuel flow to compensate for the shift, which minimized the overall fuel savings. I found little devices that could be inserted into the system and fool the computer into not registering or adjusting for the oxygen I was adding.

Though this fix did, indeed, result in better fuel savings, the device was still impossible to manage. So, I had to settle for the 10–20 percent, which was still far better than nothing.

After I started looking into ways to commercialize the device, I realized: Despite the marked increases in fuel optimization, it was just too complicated for the average individual or consumer for it to ever be commercialized in the market. You had to manually mix all of these different chemicals together and add in just the right amounts of distilled water and baking soda to make it work. You had to remember to turn some switches on at certain points to activate the device and turn other switches off at other times to prevent it from producing hydrogen all night long.

It was all just too clunky, too cumbersome, too complicated for consumers. Here I was, the lifetime tinkerer and experienced engineer who was committed to this device's success, and it was even too complicated for me. Nobody was ever going to carry all of these different chemicals and components around in their trunks and go through the trouble of manually measuring them out, mixing them together and managing the device for a mere 10–20 percent improvement.

There was also a myriad of other electrical issues that arose during testing, including problems stemming from high current and high heat. The system would just get hotter and hotter as the current increased, so I had to find another controller that would regulate the voltage (i.e., turn it on and off at various intervals) to prevent overheating.

I mean, at one point, while I was experimenting with the device on a friend's diesel truck, the darn thing caught fire. No joke, full blaze. It was my fault; I had connected the battery post incorrectly and failed to ground the unit properly, but bottom line, the thing was on fire, and you couldn't even use water to put it out. I had to smother it with my shirt. I remember almost laughing (or crying) at the absurdity of it all. Here we had water on fire—and water is not something that normally catches fire.

Though this fluke outcome (that I'm honestly happy happened) is not entirely unheard of in the process of mixing hydrogen and oxygen together while circulating in water, it would also pose some pretty serious problems when it comes to marketing. In fact, any ethical person would know that the very possibility of a misstep like that one means that the product is not (and never will be) marketable. I needed to find a new solution for these problems, or this venture would be over.

Cut to 2010 and another late night spent researching some alternative solutions, mechanisms, and processes online, when I was surprised to stumble on something new, and almost frustratingly simple. It was an application for making hydrogen gas that didn't require any of the mess of my current method—none of the measuring, mixing, chemicals (or fires). It was just a little clear cube made overseas by some Germans for educational purposes related to hydrogen production and conversion.

This simple system could take plain old water and split it up into separate streams of hydrogen and oxygen without any interference or effort from the user. It was beautiful. And it absolutely devastated me. I was sure that this discovery could only mean that somewhere, somebody else had already solved the problem I was working so passionately to solve for years now, and had done so faster than I could.

I soon realized that, while my particular use for the device would be totally different, it could produce almost precisely the same amount of hydrogen that my monster of an uncommercializable machine was making (talk about a coincidence/blessing). I'd just need to build up and around the box a bit to make it work within the context of fuel optimization. I'd have to configure a water tank to the cube-shaped device and tie all of these other pieces together in order to push the hydrogen it produced into the air intake on my car. And I smiled. I was decidedly relieved to learn that I wasn't yet out of the race.

I scrapped the old beastly device from before and immediately resumed my garage tinkering—this time, with the new model in hand. At the end of each at-home test, I became convinced that I was seeing some improvement. It appeared to actually work. Repeatedly. Without any fires. Once I'd fine-tuned my jimmy rigging of the cube well enough to be relatively confident that it would continue to work, I went to a legitimate, EPA-approved engine testing lab to start testing this thing out for real.

Once I was ready to run some tests on a dynamometer, I decided to invest in some space at the engine testing lab. I remember talking to the owner of the place, who said to me, "Well, Don, what makes you think this thing is actually going to work? I ask because we've had countless people come in here with their hydrogen devices, and none of them ever work. And you're going to spend a lot of money trying to get the thing to work. And I don't want you to be mad at me if your device doesn't work like all the rest of them." His trepidation inspired me and concerned me. However, I paid the man and made a plan.

My brother went in with me for the very first test. I'd set up equipment for two different experiments I wanted to run that day because I didn't yet know exactly how much hydrogen would be required for the cube to work with the size of the engines we were testing on. Naturally, I assumed "the more hydrogen, the better!" And went to work on the two-cube setup.

Unfortunately, those test results were absolutely horrible. I mean, they were abandon-ship bad. I think I saw maybe a 1 percent overall improvement in fuel economy, which I knew nobody would ever buy into. I remember feeling totally dejected looking at those dismal test results.

Now, this is the point where most people (including me) would probably change course, or take a breather, or go to sleep. Fortunately, I had one more experiment to run (I came there with two, remember). "OK, two cubes didn't work," I said to my brother. "Forget two. Let's test just one now. I mean, we're here, it's paid for, and I didn't make dinner plans. Let's see what one cube can do."

I flipped the test switch—partially convinced that more hydrogen is always better than less hydrogen, and therefore I was doomed—and I waited quietly for the results.

The report showed a 25 percent improvement in fuel economy. "What the?!" I sputtered. And that's when it dawned on me. You don't need as much hydrogen as you may think you need. After all, nobody had ever seen those kinds of savings before. I know they definitely hadn't seen it at this lab before, either. It was a shockingly small amount of hydrogen. So small that I am sure no other tester would have dared to show up with such a ridiculously small amount of hydrogen to test. It was totally counterintuitive to even think something like that would work.

Fortunately for me, one of my cute little cubes just **happen** to produce the exact amount to show an improvement for the relatively large engine on the vehicle I just happen to rent for the day. What a fortunate combination of events. But now, I knew it could work. I just needed to build something around it to make it much simpler to use.

That said, the device wasn't without its own faults (namely, how remarkably fragile it was) and the results were less consistent than I'd have liked on a variety of different engines. Such is the nature of engineering and experimentation—and an excellent reason to start grabbing a stiff drink after the lab. I found that the larger engines, like the one in the Chevy Suburban I tested that first day, could pretty steadily reach 20–25 percent fuel savings, but the smaller engines, like a Toyota Corolla, didn't see much marked improvements. When a car is already getting fifty miles to the gallon, I guess it's unfair to ask it to work too much harder. As I can recall, 4–5 percent.

I remember being disappointed a lot and spending several nights at home staring down at the variety of test results, thinking, "Why the heck am I being led down this road?!"

But then, one day, on my way back from the engine testing lab, it occurred to me that maybe I was overlooking a crucial element: my cube-produced oxygen. We like oxygen on earth. Earth could always use more oxygen. I remember reading something late one night that "the more the amount of oxygen, the cooler the planet." At the time, I thought, "That's interesting."

At this point, I was only testing gasoline engines. After one of my series of experiments adjusting the volume of hydrogen to be used to different engine sizes, which was really leading me nowhere, the technician asked me if I wanted to try diesel engines. "Sure, why not," I responded.

So that is how I moved to the diesel engine world and testing therein. It wasn't exactly a bed of roses in the diesel world either. One day, after another discouragingly round of test results (I was still looking for fuel economy), the technician turned to me and said, almost as a condolence to my visible disappointment, "By the way, the test report says that this thing reduces the particulate matter of the engine by almost 50 percent." My response: "OK, cool. ***What the heck is particulate matter?***" He then said something like, "I don't really know; it's just something we measure here." However, that resonated in my mind: "Cut particulate matter by almost 50 percent?" That's probably a good thing, right?

I rushed home to do some research, only to discover that particulate matter is, by definition, unburned fuel. Now, not a single gas or diesel engine on the planet can ever burn 100 percent of its fuel. It's impossible because the process happens too quickly. Diesel engines are actually worse culprits of particulate matter creation, but both types of engines churn this stuff out. After it's created, the particulate matter is thrust out of the exhaust pipe and into the air that we are breathing. It falls onto our surfaces. Heck, it can even sometimes land in our food. It's also worth noting that particulate matter is positively toxic to living, breathing life forms. It is poisonous to human beings, and to the environment. Again, we'll dig into this more later, but suffice it to say—particulate matter causes millions and millions of premature, preventable deaths each year. And we had a contraption that could cut particulate matter creation by half.

I sat there, mouth agape, as I combed through the statistics. I remember thinking, "Wow. Maybe this is the point. Perhaps particulate matter is more important than fuel economy."

Now, at the time, I still didn't know how particulate matter related to black carbon and black carbon's role in the devastation of climate change. I didn't find that out until years later.

But suddenly, it was clear that particulate matter became the new target. The new purpose. The new mission. Everything led us to that moment, and now, without knowing it, we've spent the past decade fine-tuning an extraordinarily simple solution for reversing climate change. But in order to see it, we had to step back, release our anticipated outcomes and remain open to solutions we couldn't see from the start.

As I said, I've been tinkering my whole life by taking things apart and sometimes putting them back together. I would always approach a problem through deconstruction and simplification. I'd always be more curious about learning how and why something worked the way it did than to immediately try and contrive a way to improve it. I'd always liked to seek out solutions that could work naturally and intuitively, even if that meant being wrong about their greatest applications in the end. I'd just tear something apart, tinker around within it, fine-tune its functions, and finally put it back together until it was exactly what it needed to be in order to meet its best use. I did it in the 1990s with the accounting software that I thought was too complicated at the time for the processes (and the people) it was created to serve.

As I look back, I did it with a few custom database applications I developed for several clients in the early 2000s that still use those applications to this very day, and that continue to work perfectly for them. I did it with my monstrous first fuel-saving machine that I didn't know was doomed from the start because of its inherent complications. Lastly, I did it with those cute little jimmy-rigged cubes, now greatly modified and officially patented and licensed as our LeefH2 device, which will play a pivotal role in reversing the incredibly complex, global behemoth that is climate change. It's almost comical how much luck and grief and guts have gone into this process, but it always—always—comes right back down to simplifying solutions.

And that is where we all must begin.

I didn't choose this. It's as if it chose me. This accidental outcome of one of the several fuel-optimizing, makeshift devices that I'd created in my garage to save a few bucks on gas. This unanticipated side effect of years of research and development, testing and failing, only to find out that the true solution I'd never thought to think of was there all along. The future remedy for our rapidly weakening climate on this hot rock we inhabit while it spirals through space. This cure. It chose me. It wasn't about saving the world when I started. It was just about saving some gas. How simple is that?

EPILOGUE

The Somebody Else Principle

I must confess, I never expected to write a book about anything, nor did I ever think I might ever have something really important to share with the world. I always felt that I had something inside of me that needed to be said one day, but I didn't have a clue what it might be. I've always felt that I would like to leave something or say something that would make this place (the world) a better place. Just like every last one of you that are reading this book … or that skipped to the end to read the epilogue because it's much shorter.

Well now I get to express something that I have only explained to a handful of people who I know that I have always thought was pretty profound (but possibly very confusing). I am pretty sure I made this up. I don't think I heard anyone else say it and then decided it was something that came out of me. But the next few words and paragraphs will describe a concept that is very true. It's called the "Somebody Else Principle." It is only tangentially related to climate change, but it has to do with the power of all of us. It goes like this:

"Everybody is somebody else to somebody else."

"Nobody is somebody else to themselves (or their family and friends)."

What this means is that every person sees other people (somebody else) as being uniquely capable of doing great (or bad) things. It's always "somebody else" that makes something significant happen or is associated with something significant. Take Barack Obama, Joe Biden, or Oprah Winfrey, for example (the list is endless, by the way, because it extends to celebrities, athletes, politicians, etc.), they are clearly "somebody else" in the minds of everyone because of the impact that they have had in society on whatever it is they have done (or not done).

In the minds of probably every single person on the planet, those people are who they are because they are indeed "somebody else."

Barack Obama is definitely somebody else. He was the first black president, for Christ's sake. But I promise you Barack has never seen himself as a somebody else, and I can guarantee Michelle Obama knows he's not. And I don't know Barack or Michelle to be able to say that he is just Barack to himself and just Barack (or Barry) to Michelle, but I know that that is true. I know that he is just Barack or Barry to Michelle because nobody is somebody else to their friends and family, but clearly, he is somebody else to everybody else.

But everybody is somebody else to somebody else. Which means that all the things that the somebody elses of the world have done or can do are things that we all can do because you and me both are automatically somebody else (to somebody else), and it's always somebody else that does stuff.

I should hasten to reiterate, though, that "nobody is somebody else to themselves (or their family and friends)."

No non-egotistical person ever sees THEMSELVES as anything but themselves. Never somebody else. That's because we know all of our flaws to the last painstaking degree. We know all the dumb things we have done or the lies we've told (hopefully, small insignificant ones, but we know them all). Therefore we are all just us to us and to our friends that know us.

This may sound like one of the most convoluted things you've ever heard of, but most people don't realize it or even think about it and to be honest, I am just enjoying finally telling it.

Just like you are somebody else to me, I am somebody else to you. It's always the somebody elses that make things happen in this world, but guess what, you are automatically somebody else just because you are here, which means that you can and will be capable of doing everything that the somebody elses of the world do since you are already a somebody else, because everyone is.

I am still not somebody else to myself (and never will be), but I can promise you I am definitely now somebody else to most of you. I still won't be somebody else to my sisters and brothers, friends and kids, because I am just Donnie, Don, or Dad to them ... just like you are just you to your friends and family and will NEVER be somebody else to them just like you will never be someone else to yourself because you know too many of your own flaws and shortcomings.

You (just like me) can vividly remember all of the stupid things you've done that will NEVER qualify yourself as being a somebody else (to you). But none of those things will prevent others from looking at you as somebody else. Why is that? Because you already are ... because everyone is. Which mean that you can accomplish anything on earth that you want to do because the somebody elses always do, and you are somebody else, because ...

Everybody is somebody else to somebody else.

So, continue to try to do the right things, try to be a blessing to everyone you meet, and try to follow the path that you feel in your heart that you have been given. You will then become the somebody else who you never thought you would become but that you already were because everybody is.

And, since all of you somebody elses of the world now know that you are capable of doing great things, consider helping us to fight some black carbon and climate change.

Notes

Introduction

1. *Take the Climate & Clean Air Coalition* https://www.ccacoalition.org/en/slcps/black-carbon

Chapter 1

2. *It has been estimated that, by the end of that year,* https://www.fire.ca.gov/stats-events/

3. *California's nightmarish 2020 fire season accounted for* https://fas.org/sgp/crs/misc/IF10244.pdf

4. *If we want to expand our lens to a more global view, https:*//abcnews.go.com/International/amazon-rainforest-lost-area-size-israel-2020/story?id=75683477

5. *These caused a staggering loss* https://www.aljazeera.com/news/2020/7/28/nearly-3-billion-animals-killed-or-displaced-by-australia-fires

6. *Despite the fact that yes, atmospheric CO2 levels* https://www.climate.gov/news-features/understanding-climate/climate-change-atmospheric-carbon-dioxide

7. *According to the Environmental Protection Agency (EPA)* https://www.epa.gov/pm-pollution/health-and-environmental-effects-particulate-matter-pm

8. *Beyond these more established and understood* https://www.apa.org/monitor/2012/07-08/smog

9. *According to the World Health Organization (WHO)* https://www.seas.harvard.edu/news/2021/02/deaths-fossil-fuel-emissions-higher-previously-thought

10. *According to another study recently released by Harvard* https://projects.iq.harvard.edu/covid-pm

11. *While reviewing the November study's findings* https://www.nytimes.com/2020/04/07/climate/air-pollution-coronavirus-covid.html

12. *"Black carbon is the material that burns in an orange flame,"* https://www.ncbi.nlm.nih.gov/pmc/articles/PMC3080958/

13. *While we've seen the majority of estimates* https://www.ncbi.nlm.nih.gov/pmc/articles/PMC3080958/

14. *As David Roberts, climate change journalist, puts it in a piece for Vox:* https://www.vox.com/energy-and-environment/2019/9/4/20829431/climate-change-carbon-capture-utilization-sequestration-ccu-ccs

Chapter 2

15. *In a 2001 article she penned for The American Prospect,* https://prospect.org/health/way-won-america-s-economic-breakthrough-world-war-ii/

16. *For example, when William Hart of Chautauqua* https://buffalonews.com/business/local/

nation-s-first-gas-well-was-dug-in-western-new-york/article_be62a436-6b68-552e-be33-a310addfd7e7.html

17. *And when, almost thirty years later, Samuel Kier first*
https://aoghs.org/petroleum-pioneers/american-oil-history/

18. *And when, almost thirty years later, Samuel Kier first*
https://www.post-gazette.com/life/lifestyle/2009/12/10/Let-s-Learn-About-Samuel-Kier/stories/200912100537

19. *Bissell, a lawyer, ended up forming*
https://aoghs.org/petroleum-pioneers/american-oil-history/

20. *Owen put it more precisely when he wrote,*
https://en.wikipedia.org/wiki/History_of_the_petroleum_industry_in_the_United_States#cite_ref-5

21. *They crafted the commercial blueprint*
https://www.investopedia.com/ask/answers/030915/what-percentage-global-economy-comprised-oil-gas-drilling-sector.asp

22. *According to Jonathan Clark's 2019 exposé article for Medium:*
https://medium.com/@jonathanusa/everything-you-know-about-recycling-is-wrong-well-most-everything-f348b4ee00fe

Chapter 3

23. *Sure, the California governor did just pass a landmark*
https://www.gov.ca.gov/wp-content/uploads/2020/09/9.23.20-EO-N-79-20-Climate.pdf

24. *According to a 2018 report published by the*
https://www.ucsusa.org/resources/cars-trucks-buses-and-air-pollution

25. *According to an article on Inside Climate News:*
https://insideclimatenews.org/news/31102018/truck-pollution-emission-controls-failed-health-climate-change-black-carbon-soot/

26. *According to their study:*
https://eos.org/articles/tailpipe-study-newer-trucks-emit-more-black-carbon

27. *Well, when we look more specifically at pollution*
https://www.ccacoalition.org/en/initiatives/heavy-duty-vehicles

28. *According to the Climate & Clean Air Coalition,*
https://www.ccacoalition.org/en/activity/soot-free-urban-bus-fleets

29. *According to Ian Simm, CEO of Impax Asset Management*
https://www.barrons.com/articles/vimeo-is-poised-to-be-a-public-company-revenue-is-up-57-51620245106

30. *In fact, two of the engineers who worked on one of Google's*
https://spectrum.ieee.org/energy/renewables/what-it-would-really-take-to-reverse-climate-change

31. *But I don't have to break it to you myself,*
https://www.bseec.org/renewable_energy_depends_on_fossil_fuels

32. *According to a 2017 report released by Lockheed Martin,*
https://www.lockheedmartin.com/content/dam/lockheed-martin/eo/documents/webt/transporting-wind-turbine-blades.pdf

Chapter 4

33. *The capacity for storing and optimizing hydrogen*
https://www.wsj.com/articles/japans-big-bet-on-hydrogen-could-revolutionize-the-energy-market-11623607695

34. *One report highlighted the faster charging times*
https://www.twi-global.com/technical-knowledge/faqs/what-are-the-pros-and-cons-of-hydrogen-fuel-cells

35. *According to The Essential Daily Briefing:*
https://inews.co.uk/news/long-reads/cargo-container-shipping-carbon-pollution

36. *A duo of reporters from GreenBiz*
https://www.greenbiz.com/article/truth-about-hydrogen-latest-trendiest-low-carbon-solution

Chapter 5

37. *We've been through the gamut already in the course of this book,*
https://www.forbes.com/sites/rogerpielke/2020/03/10/every-day-10000-people-die-due-to-air-pollution-from-fossil-fuels/?sh=652158272b6a

38. *That solution, our LeefH2, is real*
https://www.hnogreenfuels.com/technology

Chapter 6

39. *According to recent research released for the year 2020*
https://www.eia.gov/tools/faqs/faq.php?id=427&t=2

40. *On the flip side, however, as far as domestic energy*
https://www.eia.gov/energyexplained/us-energy-facts/

41. *When we see our global population hovering*
https://www.livescience.com/global-population.html

42. *According to Christopher Joyce, an NPR* https://www.npr.org/2019/07/09/735848489/plastic-has-a-big-carbon-footprint-but-that-isnt-the-whole-story

43. *To paint you a clearer picture in keeping with our theme,* https://stanfordmag.org/contents/the-link-between-plastic-use-and-climate-change-nitty-gritty

44. *According to Arcadia, "Coal plants* https://blog.arcadia.com/15-key-facts-statistics-power-plant-pollution/\

45. *And still, in the estimation of Caroll Muffett* https://www.npr.org/2019/07/09/735848489/plastic-has-a-big-carbon-footprint-but-that-isnt-the-whole-story

46. *According to a report released by the* https://www.ifc.org/wps/wcm/connect/dfab4f4c-9247-46ed-8f35-b25fa527b636/20190919-Summary-The-Dirty-Footprint-of-the-Broken-Grid.pdf?MOD=AJPERES&CVID=mR9UXpH

47. *According to a December 2020 study released* https://environmenthalfcentury.princeton.edu/sites/g/files/toruqf331/files/2020-12/Princeton_NZA_Interim_Report_15_Dec_2020_FINAL.pdf

Chapter 7

48. *These scenarios are both sophisticated and idealistic,* https://www.livescience.com/climate-report-net-zero.html

Chapter 8

49. *But the handheld cellular phone had already been invented* https://www.aarp.org/politics-society/history/info-2018/first-cell-phone-call.html

50. *Now, remember, that cost in 1983 looks a bit closer* https://www.in2013dollars.com/us/inflation/1983?amount=4000

51. *Considering that the iPhone alone,* https://www.forbes.com/sites/ianmorris/2017/06/29/apple-has-sold-1-2-billion-iphones-worth-738-billion-in-10-years/?sh=1def61791a18